T5-BCK-811

ENERGY AND HUMAN WELFARE— A CRITICAL ANALYSIS

VOLUME III
Human Welfare: The End Use for Power

ENERGY AND HUMAN WELFARE—A CRITICAL ANALYSIS

A Selection of Papers on the Social, Technological, and Environmental Problems of Electric Power Consumption

---

Edited by
BARRY COMMONER
HOWARD BOKSENBAUM
MICHAEL CORR

Prepared for the Electric Power Task Force of the Scientists' Institute for Public Information and the Power Study Group of the American Association for the Advancement of Science Committee on Environmental Alterations

VOLUME III

Human Welfare: The End Use for Power

MACMILLAN INFORMATION
A Division of Macmillan Publishing Co., Inc.
New York

Macmillan Information
A Division of Macmillan Publishing Co., Inc.
866 Third Avenue, New York, N.Y. 10022

Collier-Macmillan Canada Ltd.

Library of Congress Catalog Card Number: 75-8992

Printed in the United States of America

printing number

1 2 3 4 5 6 7 8 9 10

Library of Congress Cataloging in Publication Data

Main entry under title:

Human welfare.

(Energy and human welfare ; v. 3)
Includes bibliographies and index.
1. Energy consumption--United States--Addresses, essays, lectures. I. Commoner, Barry, 1917- II. Boksenbaum, Howard. III. Corr, Michael. IV. Scientists' Institute for Public Information. Electric Power Task Force. V. American Association for the Advancement of Science. Committee on Environmental Alterations. Power Study Group. VI. Series.

ISBN 0-02-468440-6

Contents

STATEMENT OF THE BOARD OF DIRECTORS OF THE AMERICAN ASSOCIATION FOR THE ADVANCEMENT OF SCIENCE

The following expression of the Board's Statement of Policy concerning AAAS Committee Activities and Reports: "Responsibility for statements of fact and expressions of opinion contained in this report rests with the committee that prepared it.* The AAAS Board of Directors, in accordance with Association policy and without passing judgment on the views expressed, has approved its publication as a contribution to the discussion of an important issue."

*Relevant to this statement of policy, the Committee on Environmental Alterations commissioned the Power Study Working Group to select the authors for the preparation of this report. The contributors to the report are responsible for the statements of fact and expressions of opinion in their respective papers.

Members of the AAAS-CEA who participated in organizing these volumes and their affiliation at that time:

Barry Commoner, Washington University
Donald Aitken, San Jose State College
F. Herbert Borman, Yale University
Theodore Byerly, U.S. Department of Agriculture
William Capron, Harvard University
H. Jack Geiger, SUNY at Stony Brook
Oscar Harkavy, Ford Foundation
Walter Modell, College of Medicine, Cornell University
Arthur Squires, City University of New York

STATEMENT OF THE COMMITTEE ON ENVIRONMENTAL ALTERATIONS

It has become clear that the majority of the people of the United States are working to stem the escalation of environmental problems resulting from the incessant expansion of industrial technology. It is this Committee's purpose to provide the public with an assessment of the environmental consequences of our technology.

Energy-related environmental problems and their associated social issues are particularly pressing since the energy industry and, in particular, the electric power industry, appear to be a major force in our economy. Lead times for major decisions in the electric power industry are of the order of 10 to 25 years, which makes it difficult for the public, the news media, and policy makers to engage in timely, well-informed debate over the crucial environmental and social problems created by increasing energy consumption.

To help fill the need for documentation on these problems, the Committee on Environmental Alterations, in collaboration with the Scientists' Institute for Public Information, established a Task Force to prepare a report on the environmental impact of power production and possible means of alleviating adverse effects. One result of the work of this Task Force has been the preparation of these three volumes. In the view of the Committee, this is a useful contribution to our understanding of this complex and urgent problem.

We recognize that each of the authors of the separate papers that make up the main part of the report have strong, well-developed views on their subjects which may not be shared by their colleagues. This is to be expected in a field which is still poorly understood. Where sharp differences of approach to a given problem exist, the report includes, if only in brief form, some material indicative of the divergence in views.

The authors of each article take full responsibility for their contributions and, on its part, the Committee on Environmental Alterations hopes that members of the scientific community and of the public at large will find this compendium a useful contribution to an increasingly urgent subject.

The Committee wishes to express its particular appreciation to Professor Dean Abrahamson and to the Task Force members for their efforts in providing this valuable document.

Acknowledgements

The editors wish to express their gratitude to all of the authors whose cooperation was so important to the production of this volume: Special thanks go to Artis Bernard for her help in resolving questions of style. Equally important was the efficient and expert help of those who contributed to the preparation of the manuscript: the artwork was done by Dru Lipsitz and Lurline Hogsett; typing and maintenance of mail communications were handled by Gladys Yandell, Peggy Whitton, Amy Papian, Lenore Harris, Jane Murdock, Sandy Marshall, Leslie Rutlin, Pieriette Murray, Cynthia Glastris and Julie Love.

We would also like to thank J. Klarmann, K. H. Lusczynski, Michael Friedlander and Caleb A. Smith for their assistance.

In addition, we wish to acknowledge the cooperation of various publishers in granting us permission to reproduce those papers that have also appeared elsewhere. Certain chapters have appeared as articles in *Environment* magazine, and are copyrighted in the indicated year by the Scientists' Institute for Public Information:

"Lost Power," April 1972
"Bottles, Cans, and Energy," March 1972
"Getting It Together," November 1972

Preface

The production of power is only one step in a complex social process that begins with the acquisition of a natural source of energy and ends with the utilization of some part of its energy content to produce goods and services. These end products are the measure of the social value of the entire process. This output can be increased by burning more fuel or, even more effectively, by improving the efficiency with which energy is converted into the desired goods and services. Indeed, this latter approach—energy conservation—is the more desirable of the two since it reduces both the environmental and capital costs incurred in producing energy. This is the subject of this final volume.

To understand the importance of energy conservation, we need to keep in mind a relatively unfamiliar concept—that the social value of a productive enterprise cannot be measured solely by its direct economic output, or in aggregate the Gross National Product. It is a fact, after all, that the GNP is increased by many activities that have no social value: for example, an automobile crash. In order fully to assess the social value of a productive activity we need to know its effect on natural resources—such as energy and environmental quality—which in turn depends upon the technology of production. This is one of the main lessons that the environmental crisis has taught us.

We now know that the basic reasons for environmental degradation in the U.S. and all other industrial countries since World War II are drastic changes in the technology of agricultural and industrial production and transportation: We now wash our clothes in detergents instead of soap; we drink beer and soda out of throwaway containers instead of returnable ones; we use man-made nitrogen fertilizer to grow food, instead of manure or crop rotation; we wear clothes made of synthetic fibers instead of cotton or wool; we drive heavy cars with high-powered, high-compression engines instead of the lighter, low-powered, low-compression pre-war types; the railroads have been supplanted by airplanes and private cars for personal travel, and by trucks for freight movement.

All of these changes have worsened environmental degradation: When a washer-full of detergent goes down the drain it causes much more pollution than the same amount of soap; a throwaway beer bottle delivers the same amount of beer as a returnable one, but at a much higher cost in pollution and trash, since the returnable bottle is used dozens of times before it is discarded; the heavy use of chemical nitrogen fertilizer pollutes rivers and lakes, while the older agricultural methods did not; synthetic fibers, unlike natural ones, are not biodegradable, so that when discarded they are either burned—causing air pollution—or clutter up the environment forever; the modern car pollutes the air with smog and lead, while the pre-war car was smog-free and could run on unleaded gasoline; airlines and private cars produce much more air pollution per passenger-mile, and trucks produce more pollution per ton-mile than the railroads.

In every one of these cases the new, more polluting technology is also more wasteful of energy than the one it has displaced. Detergents and synthetic fibers are made out of the non-renewable store of petroleum, while natural fibers and the fat needed for soap are produced, by living plants, from carbon dioxide, water and solar energy, all renewable and freely available in the environment; it takes more energy to make a throwaway bottle than it does to wash a returnable one and use it again; the modern high-compression engine delivers fewer miles to the gallon than the pre-war engines; nitrogen fertilizer is made out of natural gas, an energy cost not incurred when manure or crop rotation is used; railroads use much less fuel per passenger mile than passenger cars or airplanes, and much less fuel per ton-mile than trucks. These changes, which are described in detail in Chapters 5 and 6, lead to a general hypothesis: the sharp changes in the technology of agricultural and industrial production and transportation in the last 25 years has intensified the impact of production on the environment and has reduced the efficiency with which energy is converted into goods and services.

Here it would be useful to recall that energy is itself of no value; it becomes a source of human satisfaction only when it is *used* to produce industrial and agricultural goods and services such as transportation and communication. This means that the energy used is equal to the amount of goods and services produced multiplied by the average amount of energy needed per unit of production. The first of these factors is a measure of the resultant social value, affluence if you like; the second factor measures the efficiency with which energy is converted to goods and services, or the *energy productivity*.

A prime example of the recent trend toward reduced energy productivity is transportation, which is discussed in Chapter 2. Trucks use four to six times more fuel per ton-mile of freight carried than railroads, and this less energy-efficient technology is displacing the more efficient one. In recent years intercity truck freight has increased at the rate of 4.7% annually while railroad freight has *decreased* 2.7% annually. A similar trend is evident in passenger transporation.

Succeeding chapters (3 and 4) show that the same trend toward reduced energy productivity is evident in two other basic segments of our economy—housing and agriculture. A particularly illuminating example of these effects is the impact of the recent energy crisis on agriculture. Part of the trend toward reduced energy productivity in agriculture was the introduction about five years ago of new combines to harvest corn. While earlier machines harvested whole ears of corn after they had dried on the stalk, the new combines removed the grain from the ear, but did so before it was fully dry. To prevent rot, the grain must be hot-air dried immediately after harvest. The fuel used for this purpose is propane, which thus must be available at a given week in the season, at a reasonable price, if the grain is not to spoil. In this way, corn production has become vulnerable to fluctuations in the propane market, and through that dependency agriculture has become powerfully linked to a very different sector of the economy—the petrochemical industry. This relationship came about because most of the propane consumption in the U.S. is divided between rural and agricultural use and the petrochemical industry, in which it is an important feedstock for the production of plastics and other synthetic materials. One immediate result of the 1973 energy crisis was the development of a "grey market" in petrochemical feedstocks; the petrochemical industry, being very heavily capitalized and totally dependent on feedstocks such as propane, bid up the price well beyond even the lax legal limits. Within months the price of propane tripled and supplies became uncertain. A similar situation has affected the price of nitrogen fertilizer, which is made of natural gas—also an important source of petrochemical feedstocks. Thus agriculture became an economic victim of our technological developments.

How can we go about the urgent task of conserving energy? One idea is that it is largely up to each individual to follow energy-saving practices and to use energy-saving goods. Although life-style can affect one's energy expenditure (see Chapter 7), a good deal also depends on the kinds of goods that are available in the market—which are so frequently wasteful of energy. It is often argued that producers sell what consumers want to buy. According to this view, the reason why soaps have been displaced by detergents, small energy-efficient cars by large gas-gulping ones, cotton, wool and other natural materials by plastics and synthetic fibers, and returnable bottles by throwaway containers, is that consumers prefer detergents, large cars, plastics, synthetic fibers and throwaway containers. It follows, from this view, that our productive system cannot be turned off its present energy-wasting course until and unless people generally change their life-styles and buy goods and services that are environmentally and energetically sound.

There is no doubt that in some measure consumer preference determines what is produced. However, the effect seems to be much too limited to account for a phenomenon as general as the one that we seek to explain. There seems to be nothing in our understanding of human behavior that could lead people, quite generally, to use precisely those goods and services that happen to worsen environmental impact and the efficiency with which energy is used. Indeed, in the case of industrial goods (as distinct from consumer goods) it is difficult to see how much a choice can be governed by the life-style of the consumer, who after all has no knowledge of what the factory uses to produce the final goods. Surely such an explanation cannot account for the displacement in factories of railroad freight by truck freight, the displacement of natural rubber drive belts by synthetic ones, or the failure to develop solar energy rather than nuclear power.

We come, then, to the problem of dealing with the energy-wasting proclivities of the productive system by altering the economic principles which govern it. This is a relatively new area of concern, and one approach to it is taken up in Chapter 8. The analysis of the relationships between the economic system and the production and use of energy is still in an early stage of development. However, the growing economic crisis is a signal that this is the issue that will in the near future come to dominate the production and use of power.

October 3, 1974 **Barry Commoner**

LEE E. ERICKSON

A Review of Forecasts for U.S. Energy Consumption in 1980 and 2000

Energy forecasting is hardly an exact science. Forecasters look at past trends in varying detail, add their own judgment as to future changes in these historic patterns, and publish their estimates. Usually heavy reliance is placed on this judgment factor, but always history is the foundation of estimates of the future.

The sources of energy in the U.S. have changed dramatically in the last century. Fuel wood was the major source of energy in 1870 comprising over 70% of the total. By the mid-1880s, coal became the dominant source of energy as the railroads and steel production expanded. Coal reached its peak of importance in the first decade of the 20th century when it contributed over 75% of the total energy used. Oil began its meteoric rise in the 1920s with the emergence of large scale automobile use and by 1950 had become the primary source of energy. Natural gas began its rapid growth after WW II with the development of high pressure pipelines and the large scale earth-moving equipment needed to construct them. By the 1960s natural gas had become the second most important energy source. Most forecasters expect that in the next 30 years nuclear energy will become a significant source of energy and the dominant source of electric power. (See Figure 1 for a graphical view of these trends.)

Throughout the 20th century, irrespective of these shifts in the sources of energy, the total energy used in the U.S. has grown at an average rate of about 3.1% annually (Table 1). The fluctuations that have occurred are primarily due to changes in the rate of general economic growth and the rate of technological change.

This report is a review of forecasts to the years 1980 and 2000 for energy use in the United States. Most of the forecasts for these years

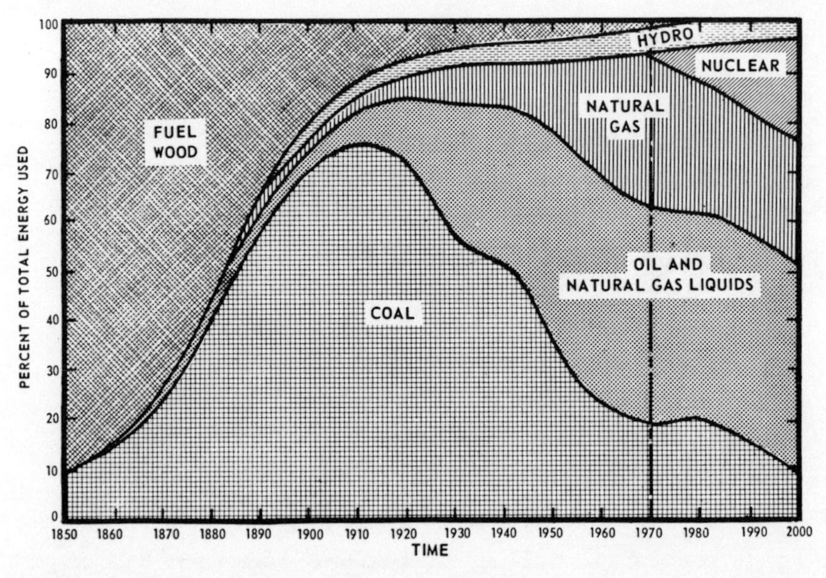

Figure 1 Sources of U.S. energy use 1850–1970 data and projections to the year 2000

Table 1 Total Energy Use—Average Annual Growth Rates for U.S.

1900–1920	4.6%
1920–1930	1.2%
1930–1940	0.7%
1940–1950	3.65%
1950–1960	2.9%
1960–1970	4.3%
1965–1970	5.0%
1960–1970	4.3%
1955–1970	3.7%
1950–1970	3.5%
1940–1970	3.3%
1930–1970	3.8%
1920–1970	2.5%
1900–1970	3.1%

Period analyzed by most forecasts 1950–1968—3.38%.

Note that while the average growth rate has been 3.1%, there have been fairly wide variations around this. Therefore, the historical base period used is critical to the forecast growth rate.

Source: Calculated from data in Bureau of Mines *Yearbooks* and *News Release* of March 9, 1971.

known to us are included, making this the most extensive review of energy forecasts to date.

FORECASTING AS A SCIENCE

Forecasting is at best an inexact science. All forecasting methods are based in some way on data concerning the historical pattern of the variable being predicted.

The simplest method is to look at the scatter of historical data points over time, "eye-ball" or "freehand" a curve through these points, and extrapolate this curve into the future. In applying this technique, semi-log graph paper is commonly used since a straight line on such paper represents a constant growth rate. Compare Figures 2 and 3 to see the effect of using semi-log paper to plot total energy use data for the U.S. and to show the average predictions discussed below. While this method is simple to apply, results obtained by different forecasters from the same data are likely to differ. This is because more or less "judgment" must be coupled with the historical data to get a forecast by this method.

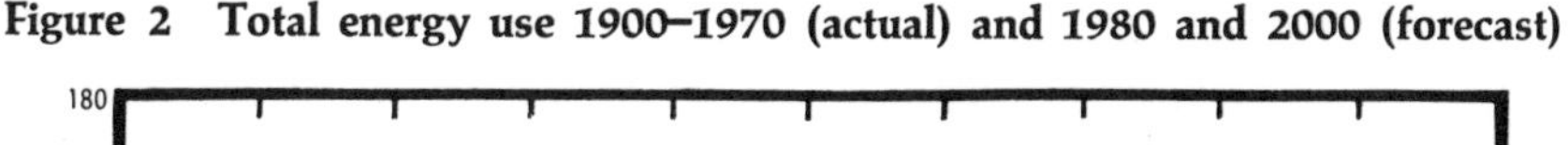

Figure 2 Total energy use 1900–1970 (actual) and 1980 and 2000 (forecast)

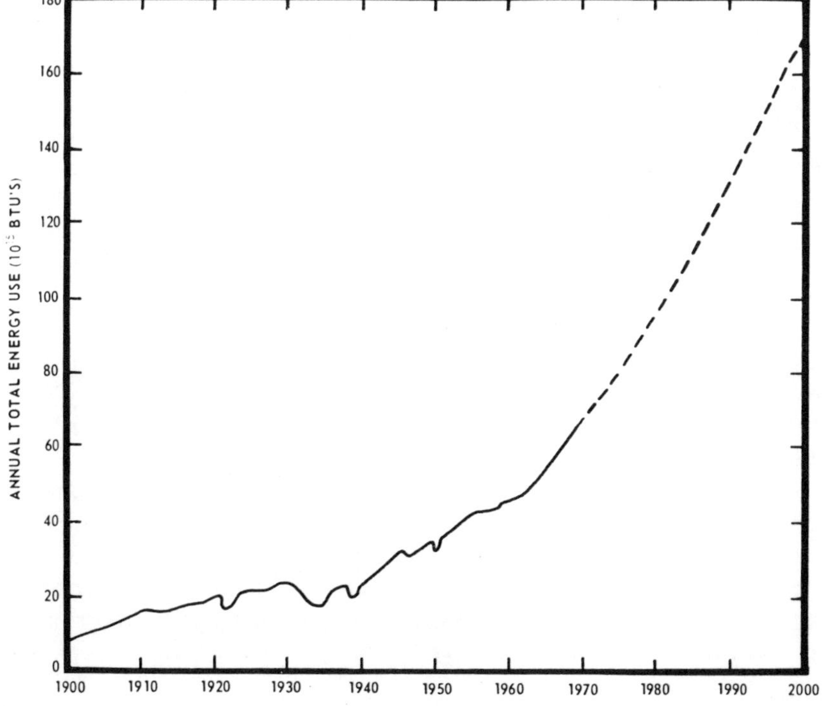

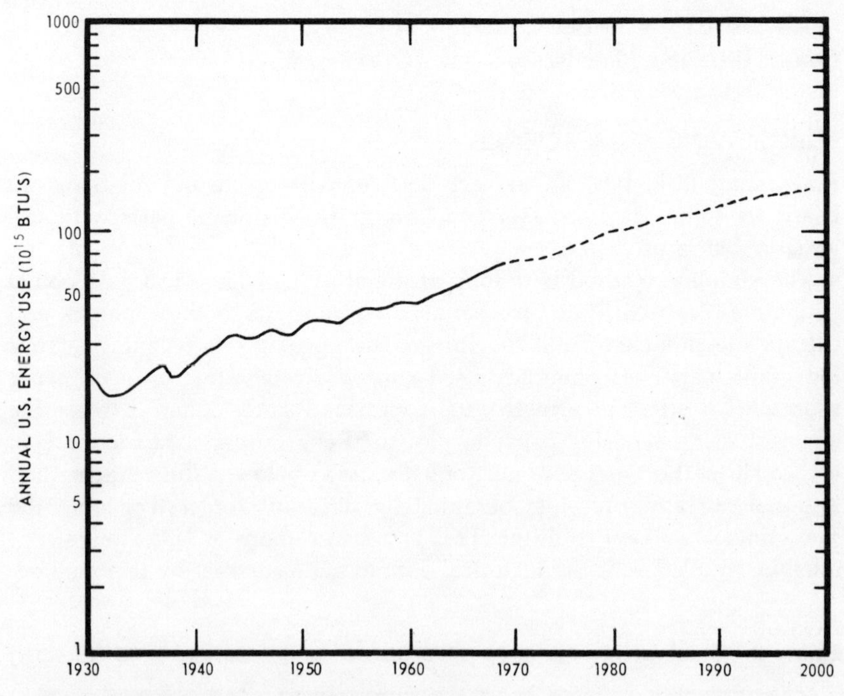

Figure 3 Energy growth is nearly constant

A variety of average or moving average methods may be used to take the judgment out of this process of drawing a curve through the data. While these methods remove the problem of different forecasters getting different results from the same data, they are still relatively primitive means of fitting a curve to the data.

The econometric or statistical method of least squares provides the best fit of a curve to the data. The curve found by this means minimizes the sum of the squares of the deviations of the data about the curve. The danger of error in the least squares method lies in misspecifying the proper form of the equation. For example, one may fit a linear equation of the form $y = a + bx$, when the data more nearly follow an exponential or semi-logarithmic equation of the form $y = ab^x$ or some more complex mathematical form. More complex econometric models can be built by forming a least squares equation from the historical relationship of the variable to be forecast to other variables and relying on projections of these other variables. But these other variables may or may not be easier to forecast and again there is the possibility that the form of the equation may be misspecified, or the time period misgauged.

All of these methods from the simplest "freehand" trend extrapolation to the most complex least squares equation are only as good as the historical data upon which they are based. Also, the further into the future one tries to forecast, the greater is the chance that conditions will arise to change historical relationships in some unforeseen way. So there is no way of being sure that a simple method which relies to some extent on the judgment of the forecaster is less accurate than a more complex econometric relationship. The accuracy in each case depends on the extent to which historical relationships change during the forecast period and the judgment of the individual forecaster in assessing the direction and magnitude of these changes. Since there was no objective basis for assuming the reliability of the forecasts reviewed they are weighted equally in obtaining the average results given. The author also developed estimates based on the average rates of growth projected for the period 1980 to 2000 and the actual use in 1970. These are shown in Figure 2 and 3 (see section results of the forecast later in chapter).

COMMON ASSUMPTIONS

The most common explicitly stated assumptions are:

1) Real GNP will grow at 4.0% annually.[1]
2) Population will grow at 1.7% annually (range 1.2–2.8%).[2]
3) Prices of fuels will remain constant relative to each other and to goods and services in general.[3]
4) Supplies of all fuels will be adequate.[4]
5) The U.S. will have no import problems.[5]

[1]This seems optimistic, although it is quite consistently assumed.

[2]Population has grown at 1.3% annually over the 1960 to 1970 decade. See Table 2 and Figure 4 for more detail on the recent history and current projections of population growth in the U.S.

[3]This is unlikely. Natural gas, oil and nuclear fuel prices will probably rise relative to other goods in the U.S. economy. Natural gas prices will be bid up as demand for this clean burning fuel exceeds supply at the existing price. Oil prices will probably rise for the same reason unless the Alaskan North Slope oil becomes available, and may rise anyway depending on political developments in South America and the Middle East and the extent of U.S. import of quotas. Nuclear fuel prices will probably rise as it will become necessary to use lower grade ores. The establishment of breeder reactors would slow this price rise.

As to the prices of fuels relative to one another, little can be said except that natural gas prices will probably continue to rise. In any case, it is hard to foresee the magnitude of these relative price changes or their quantitative effect on energy use, so they are ignored.

[4]This allows assumption (3) to hold. Note that assumptions (5) and (6) are really a part of assumption (4).

[5]This is especially important for oil after 1980. The restrictive import quota policy may be the main problem, however.

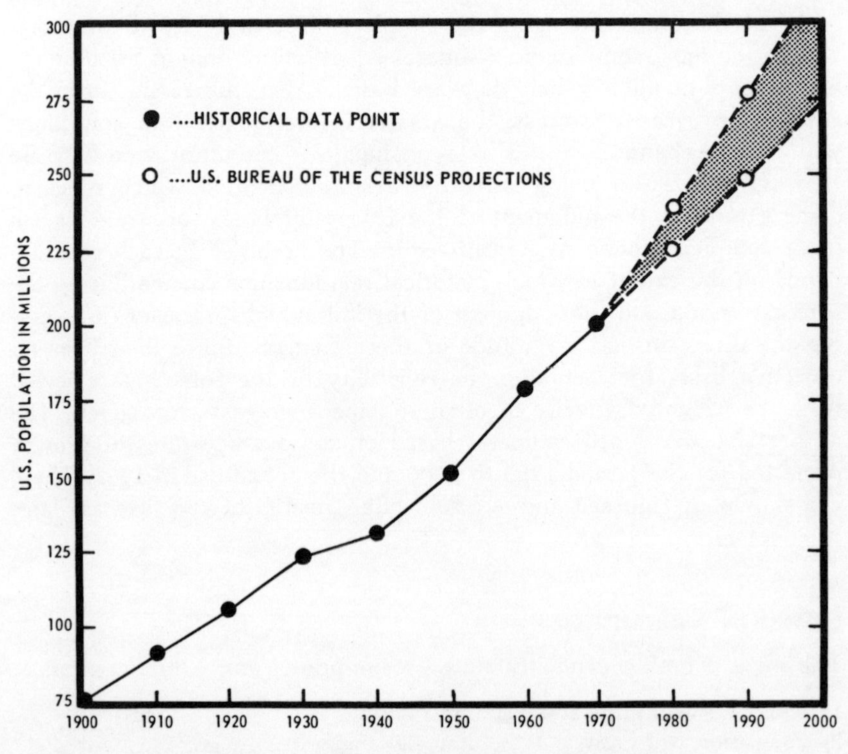

Figure 4 U.S. population 1900–1970 and range of projections

6) There will be no important environmental restrictions which are not offset by technological improvements.
7) No "revolutionary" technical change will take place, only "evolutionary" change; but nuclear power will become important after 1975 or 1980.
8) Business cycle swings will be minor.[6]
9) No major wars will occur.

Just how critically do the forecast results depend on these assumptions? Let's explore this question with the help of two simple relationships. Total energy use can be viewed both as population times per capita energy use or GNP times energy use per dollar GNP. Both of these approaches can be used to develop forecasts of total energy use. Two additional assumptions, which are commonly only implicit, are required at this point: (1) energy use per capita will increase, it is generally assumed, at about 1.3% annually, and (2) the rate of change of energy

[6]This is probably overly optimistic and could result in (1) being too high if (8) doesn't hold.

Table 2 Historical Population and Per Capita Energy Use Growth for U.S. and U.S. Bureau of the Census Population Projections

	Population (millions)	Average Annual Increase in Population During Decade	Per Capita Energy Use	Average Annual Increase in Per Capita Energy Use for Decade
1850		3.5%		
1900		1.9%	100	4.9%
1910		1.4%	161	1.4%
1920	106	1.5%	185	−0.2%
1930	123	0.68%	181	0
1940	132	1.41%	181	2.2%
1950	152	1.70%	225	
1960	180	1.32%		2.05%
1970	205	1.1–1.6%	335	
1980		0.9–1.6%		
1990				

Data Sources: Searle, 1960 and U.S.A.E.C., 1962. Bureau of Mines News Release of March 9, 1971; *Handbook of Basic Economic Statistics*, August 1971, and U.S. Bureau of the Census, *Statistical Abstract of the United States: 1971*, July 1971.

Note: The Office of Science and Technology gives a lower series of 239 and 296 million Btu/capita in 1960 and 1968 respectively. (O.S.T., 1971. *Patterns of Energy Consumption in the United States*).

use per dollar GNP will apparently fall slightly, although this may now be questionable.

Taking the most common assumptions for population and per capita energy use growth rates gives results, not too surprisingly, very close to the average values forecast of 84.3 $\times 10^{15}$ BTU for 1980 and 146.2 $\times 10^{15}$ BTU of energy use by 2000. See Table 3 for results of other assumptions. See also the section below on forecast results. Similarly taking a 4% GNP growth rate we find that most forecasts must be assuming about a 1% annual decrease in energy use per dollar GNP. If energy use per dollar GNP were constant, then a GNP growth rate of about 3% would yield the average results. For a 1% increase in the ratio, energy: GNP, and GNP growth rate of much more than 2% in real terms would be inconsistent with the average results of the forecasts as discussed below. Either a 4% GNP growth rate coupled with a 1% rate of increase in energy use per dollar GNP or a 2.5% growth rate of both population and per capita energy use would imply a rise in total energy use to nearly 300 $\times 10^{15}$ BTUs in the year 2000 compared with 68.8 $\times 10^{15}$ BTUs in 1970 and 170 $\times 10^{15}$ BTUs now expected by the turn of the century. If energy per dollar GNP continues to rise as it has since 1967, (see following section on energy/

Table 3 Combined Effects of Selected Population and Energy Use Per Capita Growth Rates on U.S. Energy Use in 1980 and 2000

Population Growth Rate	Energy Use in 1980 (2000) for Given Use Per Capita Growth Rates (10^{15} BTUs)				
	1.0	1.3	1.5	2.0	2.5
1.0	83.5 (124.2)	86.0 (136.2)	87.8 (144.1)	91.8 (166.8)	96.9 (194.0)
1.2	85.1 (132.4)	87.6 (145.0)	89.5 (153.5)	93.6 (177.4)	101.0 (203.2)
1.5	88.1 (145.0)	90.7 (158.9)	92.6 (168.3)	96.9 (194.5)	102.0 (226.0)
1.7	89.1 (153.8)	92.6 (168.3)	94.5 (178.2)	98.9 (206.0)	104.2 (239.9)
2.0	92.1 (167.8)	94.9 (184.7)	96.9 (194.5)	101.3 (225.0)	106.8 (261.7)
2.5	96.9 (194.4)	99.9 (213.0)	102.0 (225.5)	106.8 (260.5)	112.4 (303.0)
2.8	99.6 (212.0)	102.5 (232.0)	104.7 (246.0)	109.4 (284.0)	115.4 (331.0)

Source: Bureau of Mines, *News Release*, March 9, 1971 and *Handbook of Basic Economic Statistics*, August 1971, Economic Statistics Bureau of Washington, D.C.

GNP) GNP growth could well be retarded, so 300 $\times$ 10^{15} BTUs by the year 2000 is unlikely. Still this should demonstrate the importance of seemingly minor changes in assumptions. (For a more detailed look at the results of alternative assumptions see Tables 3 and 4.)

METHODOLOGY

Most energy forecasts to date have used some variation on a simple time trend. That is, the pattern of energy use over a few years is plotted and a curve is drawn "by eye." This curve is then extrapolated into the future. Different results occur depending on which years are taken as a base (Table 1) and what assumptions about future price relationships and/or technological change are used to temper the simple extrapolated results.

Many forecasters simply don't say anything about their methods. Their forecasts probably are also of the simple time trend variety, but with an added danger. The assumptions underlying the process of

Table 4 Combined Effects of Selected Rates of GNP Growth and Rates of Change of Energy Use/GNP for 1980 and 2000

	Energy Use in 1980 (2000) for Given Rates of Change of Energy Use/$GNP ($10^{15}$ BTUs)		
GNP Growth Rate	1% Decrease Annually	Constant	1% Increase Annually
2.0%	75.3 (92.2)	83.3 (124.0)	92.1 (167.3)
2.5%	79.2 (105.8)	87.5 (142.5)	96.8 (192.0)
3.0%	83.5 (123.5)	92.5 (166.5)	102.3 (224.2)
3.5%	87.4 (143.0)	96.7 (192.5)	107.0 (259.5)
4.0%	91.8 (165.0)	101.7 (222.0)	112.5 (299.5)
4.5%	96.0 (190.5)	106.5 (257.0)	118.0 (346.0)

Data Sources: Bureau of Mines *News Release*, March 9, 1971, and *Handbook of Basic Economic Statistics*, August 1971, Bureau of Economic Statistics of Washington, D.C., p. 225.

"judgment" or "expert opinion" used to temper the simple extrapolation are not spelled out fully. This sloppy way of handling quantitative variables forces the reader into the choice of either taking the forecast completely on faith or not at all.

A shining example is the statement by the National Coal Association (Senate Hearings on S.4092) that "over 200 × 10^{15} BTUs" of energy annually will be "needed" by the year 2000. At least one assumption (or lack of it) behind this statement should glare back at any economist. "Needs" or "requirements" are never absolute quantities; they are needs with respect to some (in this case unstated) goals and at some price. The unstated goals may be to allow GNP per capita to continue to grow as predicted or to allow "society as we know it" to continue or any number of other alternatives. Because energy resources, like all other resources in this world, are available only in finite, limited quantities, the competition of alternative uses for these scarce resources puts a price on these resources. Using a ton of coal to generate electricity has an "opportunity cost" of not being able to use this ton of coal to make steel. Whether or not this is expressed in dollars is purely a matter of convenience; the price or opportunity cost is still inescapably there.

So when people speak of energy "needs" or "requirements" or consumption at a given time this statement is true *at some price.* In fact almost any statement about energy consumption could be true if one supposed the right price of energy resources relative to other goods and services. Fortunately this type of guessing game is usually dropped in favor of the simplistic, but at least honest, assumption that prices of energy resources relative to each other and all other things will remain constant over the forecast period. Some forecasters admit that they have estimated what demand will be at current prices and have not looked at the more difficult question of price changes. A few forecasts assume that present price trends will continue.

Some forecasters project energy consumption by sector using a simple time trend and then aggregate these estimates to get a total energy use estimate. The most detailed of these "building block" forecasts have been made for Resources for the Future. The general procedure in *Energy in the American Economy, 1850–1975* (Schurr and Netschert, 1960) was to (1) project future output levels for specific industries, like steel, and for activities like residential space heating, (2) estimate future changes in energy use per unit of output in these uses, (3) multiply (1) and (2) to get an energy use estimate for these uses, and (4) aggregate overall uses to get a total energy use estimate. This and the later work, *Resources in America's Future: Patterns of Requirements and Availabilities, 1960–2000* (Landsberg et al., 1963) are the only works that explicitly consider factors such as the impact of insulation improvement on space heating requirements and of the rates of appliance saturation on residential electricity demand. Estimates for the use of various sources of energy were made here by the same building block procedure, although most forecasters apparently arrive at seemingly similar source estimates only after determining a figure for total energy use. These after the fact breakdowns by source have a greater danger of overlooking the extent of technical substitution which may take place between various fuels in individual uses.

If the price of one energy resource rises relative to others, this will cause some substitution of other sources of energy for this now more expensive one, increased exploration for new reserves of this resource and increased research and development (R&D) efforts into using this energy resource more efficiently. If the prices of all sources of energy rise relative to other goods and services, exploration for new reserves of all energy resources and R&D into more efficient means of using all forms of energy will be intensified and less energy use in the long run should be expected. The extent of this effect of energy prices on exploration R&D may be tempered by lack of competition in energy markets, but this is unclear.

A few forecasts use simple econometric procedures—or at least so they say. None of them reveals enough about the use of this technique for the U.S. to be evaluated. I have seen only one forecast for the U.S. which discloses the equation which was fitted, and this considered only residential demand for electricity. (FPC, 1969; Strout, 1966.) The most extensive of these studies has been done by the Bureau of Mines (1968). Extensive in this case means that many alternative assumptions about future trends of parameters like GNP, population, and available supplies were considered. "Severe" environmental restrictions were considered here; the result for 1980 was a partial switch from coal to natural gas, all other sources essentially the same and less than a one-half of one percent reduction in total energy use. Even these restrictions were considered unlikely and other sources either thought that technological change would cancel out environmental effects or ignored the issue.

At the same time forecasters have assumed only "evolutionary" technical change. This involves improvement of existing techniques but no discovery of new techniques. The distinction is difficult to understand, but it seems to be universally used. The Bureau of Mines (1968) "Energy Model" considers an exclusively natural gas fuel cell economy with no electricity produced by rotating generators and an all electric economy, but none of these "revolutionary" changes is considered likely. No one has considered such technologies as coal gasification as sufficiently likely developments for the remainder of this century to make estimates. Given present technology and fuel use, in the year 2000 an all electric economy would use 209.4×10^{15} BTUs annually, and a natural gas fuel cell economy is expected to use 106.4×10^{15} BTUs annually. (Bureau of Mines, 1968).

Relationship of Energy Use to Other Variables

Some economic and demographic variables are easier to predict than energy itself and also have a fairly stable relationship to energy use.

Indicators which have been found important in predicting energy use are:

1) Real GNP for total energy use.
2) Population and per capita real personal income for household and commercial use.
3) Index of industrial production for industrial use.
4) Real GNP for transportation use.
5) Price of electricity relative to natural gas, real personal income per capita, and population for electricity use.

There are two problems in using economic and demographic variables to project energy use. The first is that the relationship between these variables and energy use may, in time, change over rather abruptly so that even time series analysis fails to predict the change. Of course, abrupt changes in trends pose a problem with direct estimation of energy use as well. This problem is discussed for GNP below.

The second major problem is that forecasts of variables such as GNP may or may not actually be more reliable than forecasts of energy use directly (Landsberg et al., 1963). So a good understanding of the past does not necessarily lead to good predictions of the future, but it probably helps.

Per capita GNP has been closely related to per capita energy use in recent decades. Figures 5, 6, and 7 show the relationship per capita of GNP to energy use in 1952, 1961, and 1969 for a variety of nations of the world. In Canada, data for 1926–55 has given a correlation coeffi-

Figure 5 Energy consumption vs national income, 1972
Source: *Energy Requirements and Economic Growth,* Edward S. Mason

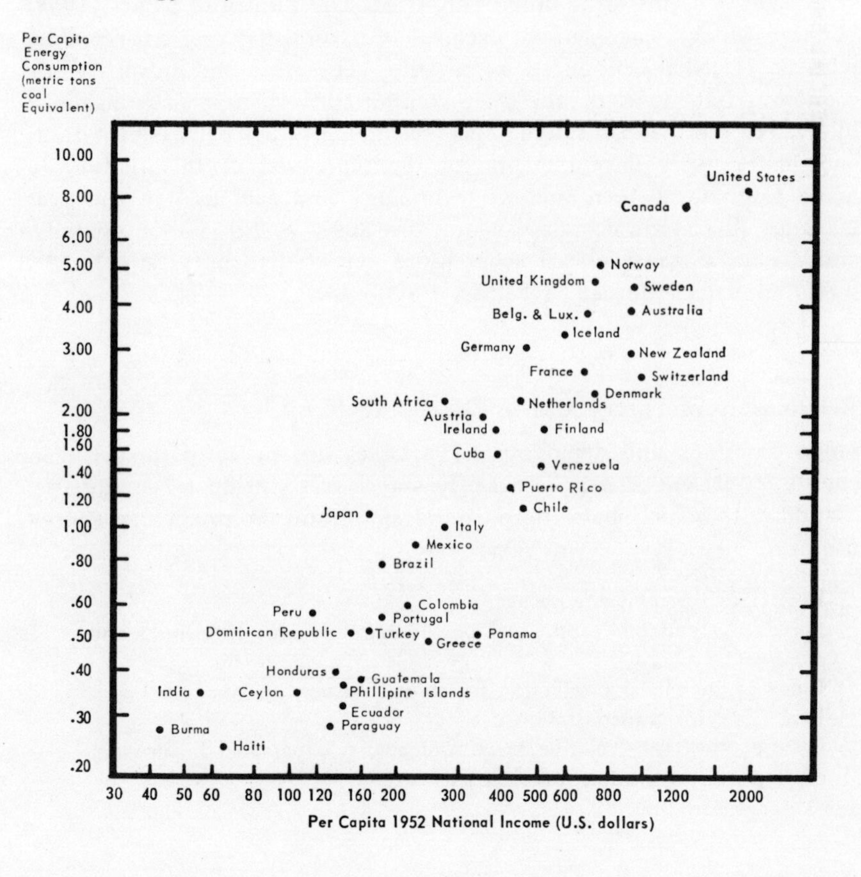

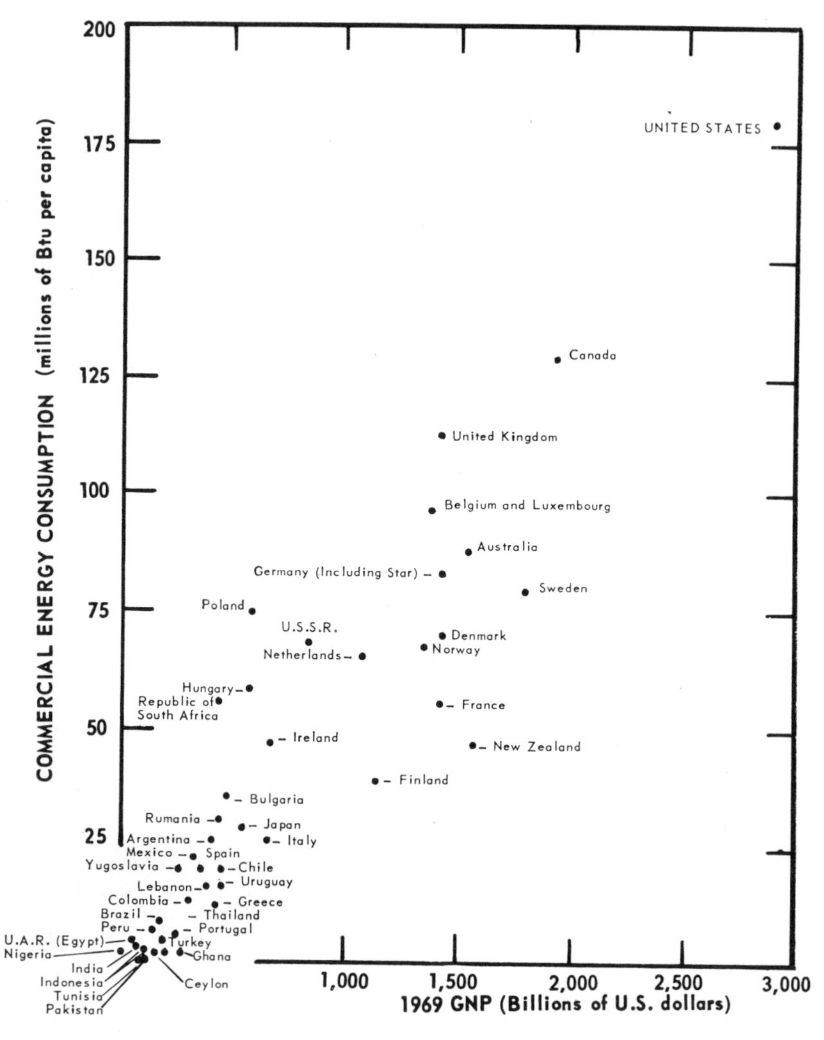

Figure 6 Per capita energy use and per capita gross national product, selected countries, 1961
Source: *Energy R&D and National Progress,* Ali Bulent Cambell

cient of 0.99 (this is a crude measure of goodness of fit of the equation of the data where 1.00 is perfect) for the relationship E = 310.54 + 0.20093 GNP where GNP is expressed in 1955 dollars (U.N.E.C.E., 1964). While such relationships have not yet been estimated for the U.S., the fact that GNP and energy use commonly go hand in hand has long been known and used in energy forecasting. However, the Bureau of Mines Energy Model uses such a projection as an upper limit or control.

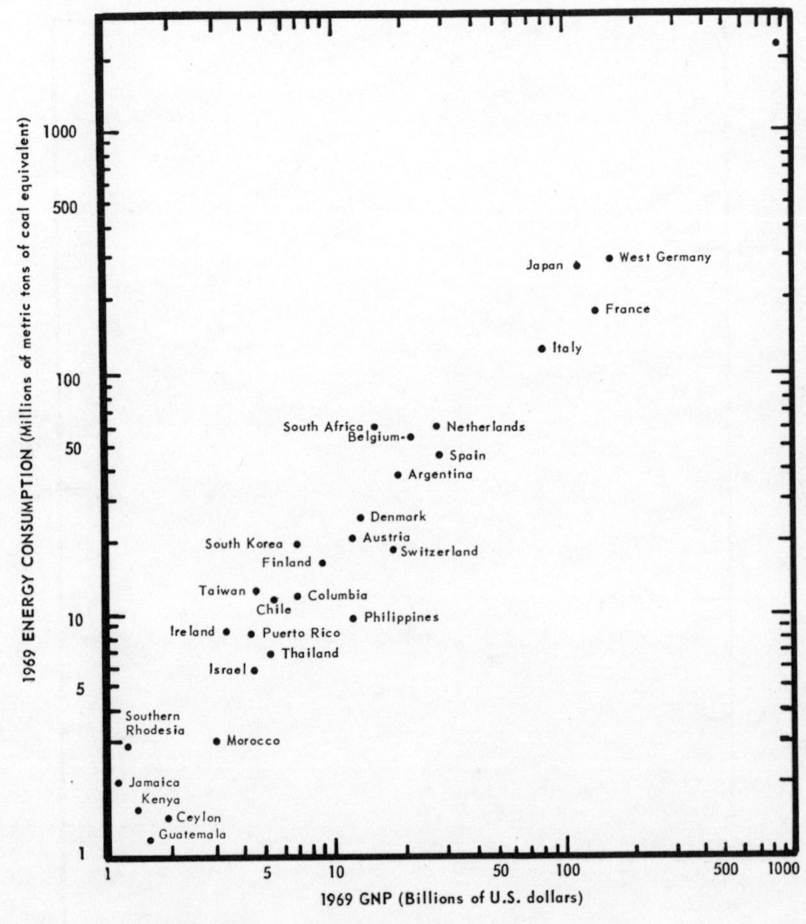

Figure 7 Energy use vs GNP selected countries, 1969

Until about 1920 (WW I) the ratio of energy to GNP (in real terms) in the U.S. was rising as energy intensive industries important for basic industrialization became established. For the last half century, this ratio has been falling as the service sector has become relatively more important (adding to GNP more than to energy use), as home heating and house size increased, and due to rapid technological improvements in the efficiency of energy use. Since about 1967 the rate of change of this ratio seems to have dramatically changed. This ratio is now rising again rather than simply becoming constant as had been expected. Energy use in the last few years has been greater than expected, and the current recession has made GNP less than expected. In 1970 energy use per GNP would have been 12% less if this change had not occurred (National Economic Research Associates, Inc., 1971). (See Figure 8.)

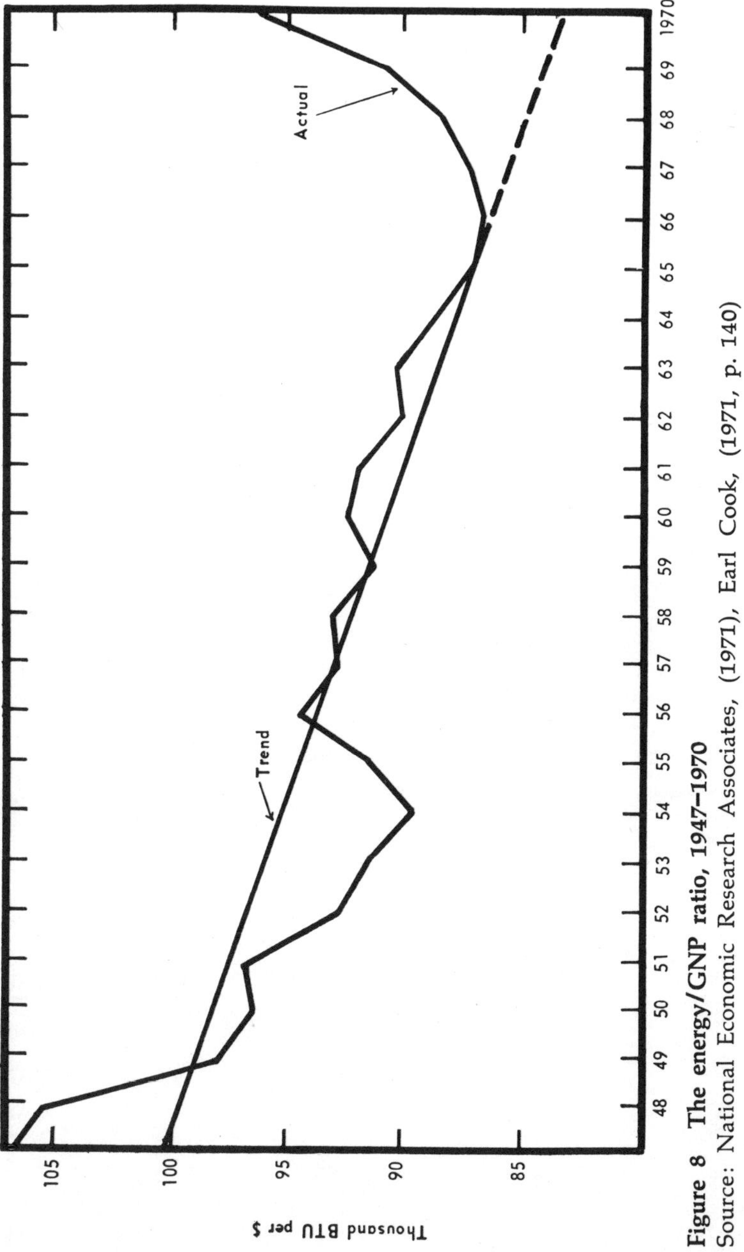

Figure 8 The energy/GNP ratio, 1947–1970
Source: National Economic Research Associates, (1971), Earl Cook, (1971, p. 140)

A preliminary analysis of this recent development by an industry source, National Economic Research Associates, cites the following reasons for this unexpected turn of events:

1) Increase in the relative importance of non-fuel use of energy resources as raw materials especially in the chemical industry.
2) Decline in the rate of improvement in the efficiency of electric power generation in fossil fuel plants. (Figure 9 shows an absolute decline for 1970.)
3) Increased importance of air conditioning and electric heat.

These factors they believe, were not by themselves enough to explain the upturn, but only a leveling off of the energy per GNP ratio.

The recent change in the trend of relationship of energy use to GNP could cause energy forecasts based on pre-1965 data to be too low if GNP forecasts are accurate, but the extent of the error depends on the number of random elements in the apparent shift.

RESULTS OF THE FORECASTS

Total Energy Use

Projections for total energy consumption in the U.S. in 1980 range from 64.5 × 10^{15} to 98.5 × 10^{15} BTUs and average 84.3 × 10^{15} BTUs. (All averages used here are simple arithmetic means.) This average total energy used annually is expected to rise to 146.2 × 10^{15} BTUs (range

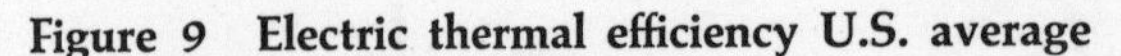

Figure 9 Electric thermal efficiency U.S. average

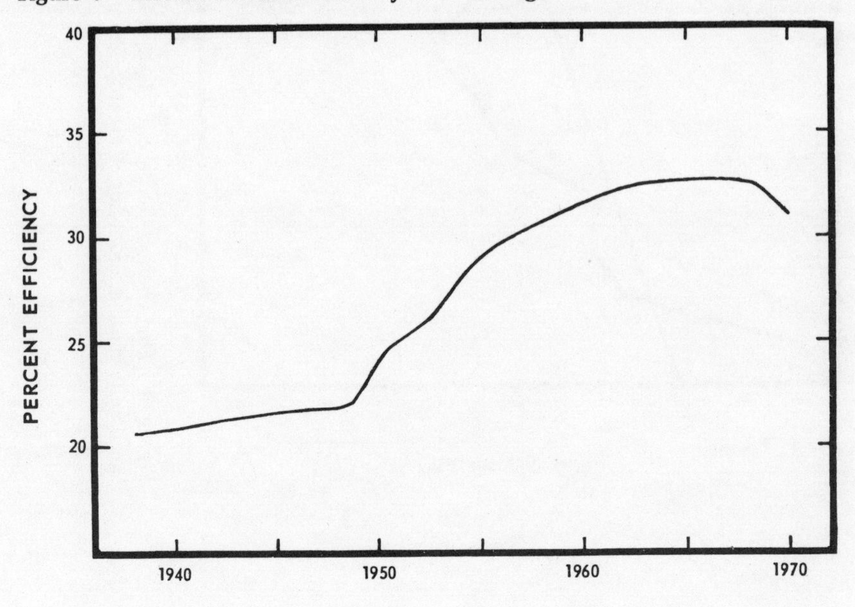

101—"over 200") by the year 2000. These forecasts were made between 1953 and 1971.

Energy use in 1970 was 68.6 $\times 10^{15}$ BTUs compared to the average prediction of 62 $\times 10^{15}$ BTUs. (For a discussion of reasons for this error, see the section on Energy per GNP above.) Given this error it seems that taking expected growth rates for the future and applying them to the 1970 base would be a reasonable up-dating procedure which would still provide estimates representative of the thought contained in these forecasts. Forecasters expect energy to climb at 3.3% (range 2.7—3.8%) annually to 1980 and then to taper off to 2.7% (range 1.5—3.8%) by the year 2000. Applying these growth rates to the 1970 base shows that energy use should be expected to be 95.3 $\times 10^{15}$ BTUs by 1980 and 170 $\times 10^{15}$ BTUs by the end of the century. (See Figures 2 and 3.) Note that these estimates still lie within the range of values actually forecast, notwithstanding the error in average 1970 forecasts.

These figures are for total fuel consumed and not energy ultimately consumed or end use energy. [For some exceptions to this almost universal rule see two works by Texas Eastern Transmission Corp. (1961, 1968)]. Because even the most modern electric generating plants achieve only about 40% efficiency in converting fuels to electricity, some energy is "lost." In addition there are transmission losses, as well as others in the economy. In 1970 this loss amounted to 11.8 $\times 10^{15}$ BTUs so that energy consumed in end uses was 57.0 $\times 10^{15}$ BTUs. Efficiency of generation of electricity has steadily improved in the past, but this trend has just about run its course, and little marginal improvement is now expected in the near future. Breeder reactors could improve this situation in the late 1980s provided they actually materialize as commercial sales. (See Figure 9 for the historical trend.) Coupled with an increasing proportion of energy being consumed in the form of electricity, this suggests that the growth rate of total end use energy will be less than the growth rate of fuel use given here. Thus we could expect about 77 $\times 10^{15}$ BTU in 1980 and 140 $\times 10^{15}$ BTU in 2000 to actually reach ultimate users assuming that they could use the resource energy as efficiently as they use the electricity. This implies 3.05% growth in total end use energy from 1970 to 2000 for a constant percentage "loss."

If the estimates for total energy use appear tenuous or confused, there is even less agreement about how the total will be distributed among consuming sectors and how it will be supplied by the various sources. Many forecasts do not even venture into these more difficult areas, so the following breakdowns are based on fewer estimates than the total figures already given. There are about two dozen forecasts of total energy use compared to about a dozen forecasts of the various individual components.

Energy Use by Sector

Household and commercial growth of energy use is expected to be less than average and declining: 2.75% (range 2.5—3.6%) until 1980 and then slowing to 2.4% (range 2.0—2.8%) during the 1980–2000 period.

While growth of industrial use will be faster than household and commercial until 1980, it will still be below average: 3.1% (range 2.7—3.8%). After 1980, it is expected to suffer a sharp drop in growth rate to 2.3% (range 1.2—3.3%) for the period of 1980–2000. Note that neither industrial nor household and commercial use figures include electricity used and that the transportation sector is not expected to use electricity.

Due to an increasing number of cars per driver as incomes increase (this will not be significant after each family unit has at least one car per driver) and an increasing proportion of the population becoming licensed drivers, the transportation sector, after an average growth rate of 3.3% to 1980, is the only sector expected to increase its growth rate thereafter (3.5% for 1980–2000).

Most forecasters expect electrical energy to increase at about 5.8% per year until 1980 with only marginal efficiency improvements. From 1980 to 2000 this rate is expected to drop to 4.8% per year, again with only marginal efficiency improvements. This is due at least partially to the fact that the efficiency of present nuclear power plants is less than that of the fossil fuel plants currently being built. For past rates of growth of fuel inputs and kw hour generated see Table 5. Again note that breeder reactors could change this trend.

A number of reasons have been suggested by various sources for the rapid, albeit decreasing, growth of electricity as compared to other sectors:

1) In the past fuel prices have increased, but due to increasing generating efficiency, electricity prices have fallen or held nearly constant. Some sources expect this trend to continue (OECD, 1969). However, most forecasters feel that generating efficiency will remain about

Table 5 U.S. Electricity Growth Rates (% Annually)

	Fuel Inputs	Electricity Generated
1950–55	5.6%	11.6%
1955–60	4.6%	6.5%
1960–65	5.7%	6.7%
1965–70	8.75%	9.6%

Source: Calculated from data in Bureau of Mines *Yearbooks* and *News Release* of March 9, 1971.

constant on average at least until breeder reactors are perfected some time in the late 1980s. For past trends in generating efficiency see Figure 9.

2) Pollution advantages: While it is true that generating plants are a source of pollution, the trend is to locate plants some distance from population centers. This removes the pollution from most of the ultimate users of the energy, who see electricity as a cleaner source of power than oil or gas. It is an open question as to whether society as a whole gains a net reduction in pollution damages, however. Fewer people suffer greater individual damage due to larger plants concentrating more pollution in more remote areas and the result is unclear. Also the ethical question of who should bear the costs of the pollution is unsettled.
3) Increasing per capita real income will mean more home appliances. Allowance is made here for some uses of electricity as yet unknown, just as TV was unknown 40 years ago.
4) Increased use of air conditioning. This use should saturate before the turn of the century.
5) Increased use of electric heat, even if it is more expensive, because it is more convenient: cleaner for the ultimate user, no noise, no furnace space, no piping, and no open flame (Landsberg, Fischman, et al., 1963). As incomes rise, people will be more concerned about such safety and convenience features. Also installation costs and costs of equipment may affect house builders and buyers more than fuel costs. After 1980, this electric heat expansion should subside as better insulation is developed and population shifts to warmer climates occur (Landsberg et al., 1963).
6) Better street lighting (Landsberg et al., 1963; U.S.A.E.C. 1962). This use should saturate also but this point was not discussed in the references cited.
7) Some allowance for the possibility of electric cars and/or mass transit by the year 2000 and some allowance for other contingencies.

As a result of the average growth rates in the various sectors, the percentage distribution of fuel use may be as follows:

	1970	1980	2000
Household and Commercial	22.5	21.2% (19.5–26.6%)	13.9% (12.0–21.2%)
Industrial	31.0	30.0% (25.8–36.8%)	24.2% (19.3–41.2%)
Transportation	21.5	21.4% (20.4–24.8%)	24.4% (24.0–27.8%)
Electric Utilities	25.0	27.4% (24.6–32.0%)	37.5% (26.0–47.0%)

Energy Use by Source

Coal use is expected to grow at about 2.7% (range 1.9—3.2%) through 1980 and only about 1% (range 0.25—2.35%) 1980–2000 as nuclear energy becomes a more important source of electricity.

Oil and natural gas liquids (NGL) are to grow at 2.9% (range 2.4—3.2%) 1970–1980 *and* 1980–2000.

Natural gas is projected to grow at 3% (range 2.1—3.6%) to 1980 and then slow to 2.6% (range 1.7—3.1%) from 1980–2000 as it becomes relatively more scarce.

Nuclear and hydropower are used only in making electricity, yet their impact on the distribution of total energy is significant. Hydropower will comprise only 2.9% of total energy in 1980 and 2.5% in 2000. By contrast, uranium is expected to produce 18.1% (range 4.9—37.6%) of the electricity generated in 1980 and 6.2% (1.4—16.8%) of total energy used then. By 2000 this new source of power is forecast to produce 52.3% of the electricity (range 37.2—60%) and comprise 19.2% (range 14.0—23.3%) of total energy used.

The resulting percentage distribution of energy use by source is roughly (especially with regard to nuclear power) as follows:

	1970 (Actual)	1980	2000
Coal	1.7%	21.3% (17.3–28.0%)	13.4% (12.0–15.0%)
Oil and NGL	42%	40.9% (39.8–41.5%)	38.4% (30.3–45.6%)
Natural Gas	32%	28.7% (27.6–30.6%)	26.6% (24.7–30.6%)
Nuclear	0.3%	6.2% (1.4–16.8%)	19.2% (14.0–26.0%)
Hydro	4%	2.9% (1.0–4.0%)	2.5% (2.1–3.0%)

The impact of nuclear power has been the biggest source of disagreement in the energy forecasts to date as evidenced by the wide range of opinion as to its use especially in 1980.

There seem to be two reasons for this:

1) Estimates of installed nuclear capacity by 1980 rose during most of the 1960s as AEC optimism began to be shared by the utility industry, so more recent forecasts tend to have somewhat higher estimates. The recent delays in construction and rapid capital cost increases for nuclear power plants have stopped the rapid increase in estimates by the AEC. (See Figure 10.)

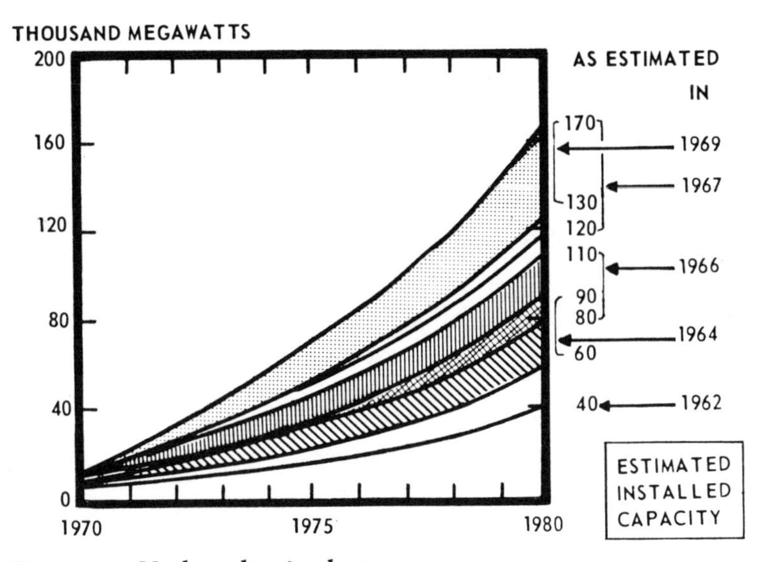

Figure 10 Nuclear electric plants
Source: The Nuclear Industry 1970, U.S. Atomic Energy Commission, U.S. Government Printing Office

2) Although the AEC estimates are widely influential, there are still some skeptics on the question of how soon the many problems with nuclear power can be worked out.

There is much more agreement that the heavily subsidized nuclear power will become an important source of energy by the end of this century providing nearly ⅕ of total energy used and over ½ of the energy used to produce electricity. No other new source of power is currently expected to become significant in the forecast period. Promising schemes which are thus being ignored include coal gasification, magnetohydrodynamics (MHD), fusion, and solar power. These techniques are not getting the R&D efforts necessary to bring their costs down to a competitive level, so they are occasionally mentioned but not included in the forecasts.

BUREAU OF MINES ENERGY MODEL

Having surveyed the general approaches and results of all the forecasts, let us now focus on one of the best studies available. The Bureau of Mines energy model was apparently developed at least five years ago, and some version of this promises to form the basis of the Bureau of Mines forecasts for the foreseeable future. (See Vogely and Morrison, 1966 and 1967; U.S. Bureau of Mines, 1968; and Morrison, 1970).

An initial estimate of 3.1% for the rate of growth of total energy use was made "using a log-log correlation of energy with GNP" (Vogely and Morrison, 1966). Apparently the following relationship was fitted from historical data for an unspecified base period (U.S. Bureau of Mines, 1968): log (GNP) = 0.7769 log (ENERGY) with a correlation coefficient of 0.99. This relationship plus the assumption that real GNP will grow at 4% annually gives the 3.1% initial estimate.

This initial estimate was then used as "an upper limit or control" (Vogely and Morrison, 1966) for the projections of energy use by sector. Relationships for the household and commercial, industrial, and transportation sectors were fitted on the basis of data from the same unspecified period. For the transport sector the relationship appears to be log (GNP) = 0.8447 log (transportation energy use) again with a correlation coefficient of 0.99. Thus a 3.6% annual growth rate for energy use in the transportation sector results (U.S. Bureau of Mines, 1968). However, the household and commercial, and industrial sectors don't follow the constant elasticity form given for total energy and the transport sector above. The form of the equations fitted which specify how the elasticities change is not apparent, but the following are consistent with the results given (U.S. Bureau of Mines, 1968):

1) log (population) = A log (household & commercial energy use)
 where A = 1.7814 for 1965–1980
 A = 0.6250 for 1980–2000

 and the correlation coefficient = 0.99 for the unspecific relationship and base period.
2) log (index of industrial production) = B log (industrial energy use)
 where B = 0.4273 for 1965–1980
 B = 0.3182 for 1980–2000

 and the correlation coefficient = 0.89 for the unspecified relationship and base period. Population is assumed to grow at 1.6% annually and the index of industrial production is assumed to grow at 4.4% per year. The resulting growth rates are as follows: Household and commercial energy use grows 2.9% annually for 1965–1980 and 1.0% for 1980–2000 excluding electricity, and 3.5% annually for 1965–1980 and 3.0% for 1980–2000 including electricity. Industrial energy use grows at 2.4% for 1965–1980 and 1.4% for 1980–2000 excluding electricity, and 2.7% for 1965–1980 and 2.1% for 1980–2000 including electricity. Electricity projections are taken from the *1964 National Power Survey* by the F.P.C. for 1965–1980 and are simply extrapolated to 2000 at the same 5.4% annual growth rate.

The resulting breakdown of energy use by primary use sector for the year 2000 is as follows (Morrison, 1970):

Household and commercial	12%
Industrial	19%
Transportation	25%
Electric Utilities	44%

The breakdown of energy use by fuel source is as follows for the year 2000 (Morrison, 1970):

Coal	12%
Natural Gas	25%
Petroleum and NGL	34%
Hydro	3%
Uranium	26%

These projections of energy use by sector seem to be consistent with a 3.3% constant annual growth rate of total energy use. Just how the "upper limit" of 3.1% works, if at all, is not explained. Also it is unclear just how this procedure is consistent with the relationship that "the basic energy model may be expressed as $E = f(X1^b, X2^c, \ldots, X13^n)$ (U.S. Bureau of Mines, 1968). In addition to not being the precise form of an equation many of the independent variables are not specified precisely enough to act as dummies let alone be quantified. For example, X7 is "environmental restrictions," X10 is "regional factors," X11 is "energy policy," and X13 is "other variables accepted as given."

Before we leave the model itself one more point which has already been mentioned deserves emphasis. The base period used for fitting the equations (however specified) has not been specified. Table 1 shows how the growth rates of energy use in the U.S. have varied in this century. In view of this, the base period chosen is really critical. Even with these criticisms, it should be remembered that this forecast was chosen for more detailed discussion because it is comparatively thorough and well formulated. That is, other forecasts have at least as many cases of omitted assumptions and unclear specifications of methodology.

One of the real virtues of the Bureau of Mines energy model is that it explores the effect of assumptions other than the ones considered most likely. The predictions of the model as presented above are 88.1×10^{15} BTUs in 1980 and 168.6×10^{15} BTUs total end use energy in 2000 which imply a 3.3% constant growth rate of energy use. Changing the GNP and population growth rates to 2.5% and 1.0% respectively lowers these predictions to 73.9×10^{15} BTUs in 1980. Raising the GNP and population growth rates to 5.5% and 2.2% respectively, changes the forecast to 104.7×10^{15} BTUs in 1980. The growth rates of total energy use vary from 2.2% to 4.5% for 1965 to 1980 depending on which of these assumptions is chosen.

Table 6 Effect of Selected Growth Rates on U.S. Energy Use for 1980 and 2000

Selected Growth Rates 1970–2000	Energy Use (10^{15} BTUs) 1980	2000
2.0%	84	124.7
2.5%	88	144.5
3.0%	92.5	167
3.3%	95.3	183
3.5%	97	193.5
4.0%	102	223
4.5%	107	257.5
5.0%	112	298

Source: Calculated from data in Bureau of Mines *News Release* of March 9, 1971.

Table 7 Effects of Selected Growth Rates on U.S. Population in 1980 and 2000[a]

Growth Rates	Population (millions) 1980	2000
1.0%	226	275
1.2%	230	293
1.5%	238	321
1.7%	243	340
2.0%	249	371
2.5%	262	430
2.8%	269	469

[a]1970 base = 204.8 million as noted in *Handbook of Basic Economic Statistics*, Economic Statistic Bureau of Washington, D.C., August 1971.

Table 8 Effects of Selected Growth Rates on U.S. Energy Use Per Capita in 1980 and 2000[a]

Growth Rate	Energy Use/Capita (10^8 BTU/Capita) 1980	2000
1.0%	3.70	4.52
1.3%	3.81	4.95
1.5%	3.89	5.24
2.0%	4.07	6.06
2.5%	4.29	7.05

[a]1970 base $= \dfrac{68.6 \times 10^{15}\ \text{BTU}}{204.8 \times 10^{6}\ \text{people}} = 3.35 \times 10^{8}\ \dfrac{\text{BTU}}{\text{capita}}$

From data in Bureau of Mines *News Release* March 9, 1971, and *Handbook of Basic Economic Statistics.*

Table 9 Effects of Selected Growth Rates on Real GNP in 1980 and 2000[a]

Growth Rate	GNP (10^{12} 1970 dollars)	
	1980	2000
2.0%	1.18	1.76
2.5%	1.24	2.02
3.0%	1.31	2.36
3.5%	1.37	2.73
4.0%	1.44	3.15
4.5%	1.51	3.64

[a]1970 base = $974.1 billion as noted in *Handbook of Basic Economic Statistics,* August 1971. Economic Statistics Bureau of Washington, D.C., p. 225.

For the period 1980 to 2000 the effects of changes in energy sources are explored. A system of natural gas fuel cells used exclusively with no intermediate electricity production would result in only 106.5×10^{15} BTUs being consumed in 2000. By contrast an economy which used only electricity except for raw materials and which used the same proportion of fuel inputs as at present would use 209.4×10^{15} BTUs in 2000, or 224.9×10^{15} BTUs if the electricity were produced entirely by coal in thermal power plants (U.S. Bureau of Mines, 1968). The most efficient all-electric system based on known technology would require 161×10^{15} BTUs in 2000; this is the MHD steam plant using coal as fuel (Morrison, 1970).

No new technologies such as MHD, fuel cells, coal gasification, or solar energy have been thoroughly considered in any forecasts because of the currently small likelihood of their adoption on a large scale during the forecast period. Federal R&D expenditures have much to do with the expectation that nuclear power including breeder reactors are expected to become important while other promising technologies are

Table 10 Effect of 1% Increase or Decrease Annually on Energy Use/$GNP for 1980 and 2000[a]

	10^4 BTU/$GNP (1970 dollars)	
	1980	2000
1% decrease annually	6.38	5.23
1% increase annually	7.81	9.50

[a]1970 base $= \dfrac{68.8 \times 10^{15}\text{ BTU}}{\$974.1 \times 10^{9}} = 7.06 \times \dfrac{10^{4}\text{ BTU}}{\$\text{GNP}}$

From data in Bureau of Mines, *News Release* March 9, 1971, and *Handbook of Basic Economic Statistics.*

expected to be largely neglected. So an important factor held constant in the forecasts reviewed here is public energy policy.

REFERENCES

American Association of Petroleum Geologists, 1969. *Future energy outlook.* 53rd Annual Meeting, Colorado School of Mines.

American Gas Association Inc. *Gas utility and pipeline industry projections 1968–1972, 1975, 1980, and 1985.* Dept. of Statistics.

Anderson, D., 1970. Ex-post evaluation of electricity demand forecasts. International Development Association, June 18.

Battelle Memorial Institute, Dec., 1969. *A review and comparison of selected United States energy forecasts.*

Cambell, Ali Bulent, 1964. *Energy R and D and national progress.* U.S. Energy Study Group, June.

———, 1966. *Energy R and D and national progress: Findings and conclusions.* U.S. Energy Study Group, Sept.

Canadian National Energy Board, 1969. *Energy supply and demand in Canada and export demand for Canadian energy 1966 to 1990.* Ottawa, Canada.

Chase Manhattan Bank, October, 1968. *Outlook for energy in the U.S.*

Committee on Interior and Insular Affairs, Sept. 1962. *Report of the national fuels and energy study group on an assessment of available information on energy in the United States.*

Dept. of the Interior. News Release, Jan. 13, 1971.

Dole, H. M., 1971. Speech at Stanford Univ., Jan. 12, 1971.

Environmental Protection Agency of the City of New York. *Toward a rational power policy: Reconciling needs for energy and environmental protection.* Apr. 1971.

Federal Power Commission (FPC), 1964. *1964 national power survey.*

———, 1969. *Methodology of load forecasting.* Report to FPC by Technical Advisory Committee on Load Forecasting Methodology for the National Power Survey.

———, 1970. *1970 national power survey.*

———, 1970. FPC annual report for 1970.

Felix, F., 1970. *Electrical World,* July 6.

Future Requirements Agency, Denver Research Institute, 1967. *Future natural gas requirements of the United States.* Univ. of Denver, vol. 2, June.

Joint Economic Committee of the Congress, Sept. 1, 1970. *The economy, energy, and the environment.*

Landsberg, Hans H., 1960. Long-range projections of energy consumption by class or type of use. *Methodology of load forecasting.*

———, and Sam H. Schurr, 1968. *Energy in the United States.*

———, et al., 1963. *Resources in America's future: Patterns of requirements and availabilities, 1960–2000.* Baltimore: Johns Hopkins Press.

MacAvoy, P. W., 1969. *Economic strategy for developing nuclear breeder reactors.* Cambridge, Mass.: MIT Press.

Mason, E. S., 1955. *Energy requirements and economic growth.* National Planning Association.

Morrison, W. E., 1970. Energy resources and national strength. Presented at the Industrial College of the Armed Forces, Oct. 6.

Nassikas, J. N., 1969. Hearings on supplies of natural gas. 91st Congress, 1st Session, November 13–14, 1969, Committee on Interior and Insular Affairs, Subcommittee on Minerals, Materials, and Fuels.

R. R. Nathan Associates, Inc., 1968. *Projections of the consumation of commodities producible on the public lands of the United States, 1980–2000.*

National Coal Association, 1970. Hearings on S. 4092 before the Subcommittee on Minerals, Materials, and Fuels of the Senate Committee on Interior and Insular Affairs, Sept. 10 and 11, 1970.

Oil and Gas Journal, 1969. Forecast of the seventies. Nov. 10.

———, Mar. 8, 1971.

Olmstead, L. M., 1970. 21st annual electrical industry forecast, *Electrical World,* Sept. 15, pp. 35–50.

Organization for Economic Co-operation and Development (OECD), 1969. *Impact of natural gas on the consumption of energy in the OECD European member countries.* Paris.

Putnam, P. C., 1953. *Energy in the future.*

Schurr, S. H., and B. C. Netschert, 1960. *Energy in the American economy 1850–1975.*

Searl, M. F., 1960. *Fossil fuels in the future.* No. TID-8209, Oct. 1960.

Sortorius and Company, Sept., 1967. *Energy in the United States, 1960–1985.*

Strout, A. M., 1966. *Technological change and United States energy consumption, 1939–1954.* Ph.D. Diss., Univ. of Chicago.

Texas Eastern Transmission Corp., 1961. *Energy and fuels in the United States 1947–1980.* P.O. Box 2521, Houston, Texas.

———, 1968. *Competition and growth in American energy markets, 1947–1985.*

U.N. Economic Commission for Europe (UNECE), 1964. Methods and principles for projecting future energy requirements.

U.S. Atomic Energy Commission, 1962. *Civilian nuclear power.*

———, 1967. *Forecast of growth of nuclear power.* WASH-1084, Dec., 1967.

———, 1970. *Civilian nuclear power: Potential nuclear power growth patterns.* WASH-1098, Dec., 1970.

———, 1971. *Forecast of growth of nuclear power.* WASH-1139, Jan., 1971.

U.S. Bureau of Mines, 1968. An energy model for the United States, featuring energy balances for the years 1947 to 1965, and projections and forecasts to the years 1980 and 2000. Info. Circular No. 8384, U.S. Dept. of the Interior.

U.S. Dept. of the Interior, Office of Oil and Gas, Jan. 1965. *An appraisal of the petroleum industry of the United States.*

———, July 1968. *United States petroleum through 1980.*

Vogely, W. A., 1962. *Patterns of energy consumption in the U.S.* Bureau of Mines.

Vogely, W. A., and W. E. Morrison, 1966. Patterns of energy consumption in the United States 1947 to 1965 and 1980 projected. Presented at the World Power Conference, Tokyo, Oct. 16–20.

———, 1967. Patterns of U.S. energy consumption to 1980. IEEE Spectrum, Sept.

D. P. GRIMMER AND K. LUSZCZYNSKI

Energy Use in Transportation[1]

Transportation now claims a large share of the energy expended in this country. As the demand for transport fuel grows and limitations on energy resources become apparent, questions arise about the best use of energy for transporation. This consideration may in fact prove to be decisive in the selection of desirable methods of transportation. The present review surveys some of the background information on energy use in transportation (Grimmer and Luszczynski, 1971). Our analysis suggests that new policies must be formulated to make more efficient use of dwindling fossil fuel reserves. Such policies seem possible on the basis of current technology, but may require substantial changes in many areas of transportation.

Energy used in all activities is derived from coal, petroleum, gas, and nuclear material, as well as from elemental or nonfuel resources such as hydropower. Other forms of energy resources, such as animate energy, indirectly derived from the sun, represent a very small fraction of the energy consumed and are not considered here.

The 1968 U.S. energy budget in Table 1 shows amounts of energy consumed in the U.S. by various sectors. Fossil fuels (anthracite, coal, lignite, natural gas, petroleum), water power, nuclear power, and electricity supplied to the individual sectors are converted into British Thermal Units (BTUs) for comparison. Things are somewhat complicated by the fact that it is necessary to distinguish between the Gross Energy Input (GEI) and the Net Energy Input (NEI), as shown in Table 1. The Gross Energy Input is equivalent to all types of commercial energy which are incorporated into the economy, whether pro-

[1]Reprinted, with permission, from *Environment*, Vol. 14, No. 3, April, 1972.

Table 1 1968 U.S. Energy Budget in QBTUs (Quadrillion BTUs)

Sector	Gross Energy Input (GEI)	Electricity Purchased (or Sold)	Net Energy Input (NEI)	Percentage of NEI
Household and commercial	13.6	2.5	16.1	30.4%
Industry	19.4	2.0	21.4	40.5%[a]
Transportation (propulsion)	15.2	0.0[b]	15.2	28.7%
Electricity, generation, utilities	14.0	(4.5)[c]	[d]	—
Totals[e]	62.4	. . .	52.9	100%

[a]Industry generated about 0.34 QBTUs of electricity (Federal Power Commission, *1969 Annual Report*).
[b]Propulsion accounted only for 0.018 QBTU or 0.4% of electricity sold.
[c]This is equal to the total electricity purchased. The energy lost in electricity generation by utilities was 9.5 QBTUs.
[d]Electricity sold (4.5 QBTUs) is added into first two entries above, and thus is not included here; the remaining 9.5 QBTUs of electricity were those lost by utilities, and thus are not included either. The NEI thus is energy available for end use by households, commercial establishments, industry, and transportation.
[e]Includes miscellaneous and unaccounted for uses (about 0.3 QBTU). Entries may not add to total because of rounding off of numbers.
Source: U.S. Bureau of Mines, *Minerals Yearbook*, p. 24, 1968.

duced domestically or imported. The Net Energy Input is the amount of the GEI minus energy lost in electricity generation and is thus approximately equal to the energy available for end use.

As shown in Table 1, energy used for transport propulsion accounts for about 29% of NEI, or total U.S. energy consumption. Fuel for automobiles, trucks, busses, and jet airplanes accounts for 81% of the propulsion energy. Further breakdown of the NEI data in the industrial sector reveals that energy used by the transportation sector is in fact considerably higher than 29%. When energy consumption for secondary-transport-related activities such as fuel refining and manufacture of transportation equipment is included, energy expended for transportation rises to a total of 38% of the NEI, as shown in Table 2.

In addition to the principal uses shown explicitly in Table 2, there are many small miscellaneous items which are difficult to evaluate because there is a lack of good data in this area.

For example, we estimate that about 0.6% of U.S. NEI goes into concrete for highways, and another 0.8% goes into motor vehicle lubricating oil. Though individually small, items such as these, when added together, represent an appreciable fraction of energy used in transportation.

Table 2 Energy for Transportation in QBTUs (Quadrillion BTUs)

Sector	Use	(QBTU)	Total QBTU	% of Net Energy Input[a]
Primary Transport (propulsion)	Automobile	7.60		
	Truck and bus	3.06		
	Jet	1.63		
	Railroads	0.72		
	Marine	0.55		
	All other prop.	1.59	15.15	29%
Secondary Transport (related activities)	Fuel refining, asphalt and road oil, energy	2.42		
	Primary metals used in transport manufacture	1.02		
	Manufacturing	0.53		
	All other secondary	1.05	5.02	9%
Total				38%

[a]See Table 1 caption for definition of Net Energy Input.

The present transport propulsion is almost entirely based on petroleum, which provides 95% of the energy; electric power supplies only about 0.1%. However, a good deal of electricity is used in the secondary-transport-related activities. Using U.S. Bureau of Mines (1968) statistics, we estimate that utility electricity use in this area accounts for about 0.9% of U.S. NEI, or about 11% of electricity purchased in 1968 (Grimmer and Luszczynski, 1971), a factor which does not show up explicitly in Tables 1 or 2.

EFFICIENCY OF TRANSPORTATION

The concept of transportation efficiency is directly related to the amount of fuel used per unit of transportation. In these terms, busses and commuter trains have transportation efficiencies several times higher than automobiles. Air passenger transport is considerably less efficient than any common mode of passenger transport on land. In freight transport, trucks and aircraft use more fuel per cargo ton-mile than watercraft, pipelines, or railroads. Freight transport by air is especially costly in terms of fuel used per ton-mile; thus air transport is very inefficient as far as fuel use is concerned.

Table 3 Efficiency of Modes of Passenger Transport

Modes	Cruise Power (H.P.)	Speed (mph)	Seat Capacity (no.)	Occupancy Assumed (%)	Transportation Efficiency (Passenger mile per gal.)	Transportation Efficiency (BTUs per passenger mile)[a]
Rail						
Fast train[b]	2,400	100	360	55	133	980
Commuter train[c]	4,000	40	1,000	50	100	1,300
Cross-country train	2,400	60	360	55	80	1,600
10-car subway train[d]	4,000	30	1,000	50	75	1,700
Road						
Large bus	200	50	43	58	125	1,000
Automobile (sedan)	50	67	4	25–50	16–32	8,100–4,100
Air						
747-jet	60,000	500	360	55	22	5,900
707-jet	28,000	500	136	62	21	6,200
STOL[e] plane (4-prop)	10,000	200	99	55	18	7,200
SST (US)	240,000	1,500	250	60	13.7	9,500
Helicopter (3 engine)	12,000	150	78	58	7.5	17,300

[a] Conversion obtained with 130,000 BTUs per gallon.
[b] 3-car, self-propelled, bi-directional double-deck, 67 tons per car.
[c] 10-car train with 2 diesel locomotives, 950 tons gross weight.
[d] New N.Y. subway train at heavy, non-rush hour traffic.
[e] Short-take-off-and-landing (see "Aircraft in the Balance," Environment, December, 1971).
Source: Rice, R. A., ASME, 70 WA/ENER-8, November 1970. Table 12.

Figures for transportation efficiency are obtained from the amount of transportation provided and the quantity of fuel used by different modes. In passenger transport the unit of transportation is taken as one passenger-mile, that is, transportation of one passenger over a distance of one mile. For freight transport the corresponding unit is one ton-mile.

Table 3 lists some representative figures for different modes of passenger transport. The transportation efficiency of a mode can be expressed in passenger-miles per gallon of fuel, or alternatively, in terms of energy expended in BTUs per passenger-mile. The lowest BTU per passenger-mile figures correspond to highest efficiencies. For example, the commuter train transports, on the average, 100 passengers for a distance of one mile for each gallon of fuel used, whereas the automobile with an occupancy of between one and two people per car transports sixteen to thiry-two people for a distance of one mile for the same amount of fuel. Large jets have efficiencies comparable to those of the automobile, while helicopters have the lowest efficiencies. Air transport efficiencies are very much lower than those of intercity busses or trains.

Pipelines, trains, and barges have transportation efficiences in excess of 200 ton-miles per gallon of fuel, while trucks transport some 50 ton-miles for each gallon.

It should be noted that, as a rule, transportation modes with poor efficiency also have high pollutant emissions per unit of transportation. For example, among the common modes of land transportation, the current internal combustion engine automobile has the highest pollutant emissions as well as the highest fuel consumption per passenger-mile. Unfortunately, devices that limit emissions from the automobile tend to lower its efficiency even more.

The distribution of fuel used and the amounts of intercity transportation provided by various modes in the mid-1960s are shown in Figure 1. Most of the intercity passenger travel is accomplished by automobiles; travel by air comes second. These two modes have relatively low transportation efficiencies, so the overall passenger transportation efficiency is low. The amount of total fuel used for passenger transport was nearly three times the total fuel used for cargo.

In 1965 the intercity cargo transport by rail accounted for about 44% of net ton-miles and used about 30% of intercity cargo fuel, while trucks transported about 22% of cargo and consumed 52% of fuel. The role of air transport was relatively small. From the point of view of fuel costs it is important to note that increase in transport by trucks and airplanes lowers the overall transportation efficiency figure.

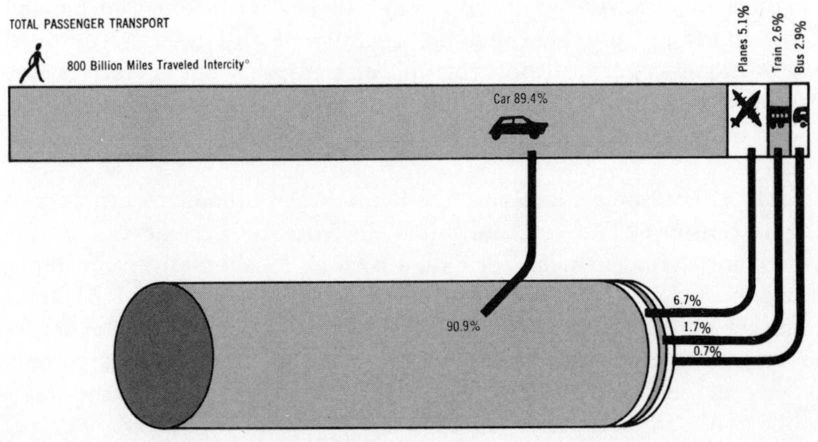

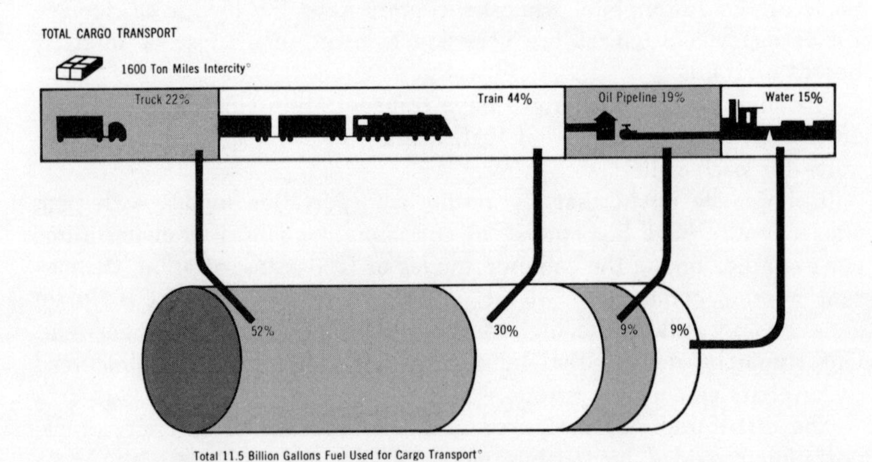

Figure 1 Fuel used for intercity transport, 1965
Source: Rice, R. A., ASME, 70 WA/ENER-8, Nov. 1970, Table 6

The increased use of various transportation modes and their fuel requirements are briefly summarized next in the context of known fuel resources (Grimmer and Luszczynski, 1971).

ENERGY USE IN TRANSPORTATION

Projections for the year 2000 are given in Table 4, which is based on U.S. Bureau of Mines (1970) statistics. According to these statistics the

Table 4 Annual Energy Consumption for 1968 and 2000 in QBTUs (Quadrillion BTUs)

Consumer	Coal		Gas		Petroleum		Hydro		Nuclear		Electricity: Gross Energy Input		Electricity: Purchased (sold)		Electricity: Net Energy Input	
	1968	2000	1968	2000	1968	2000	1968	2000	1968	2000	1968	2000	1968	2000	1968	2000
Household and Commercial	0.6		6.5	19.1	6.6	2.0					13.6	21.1	2.5	20.0	16.1	41.0
Industrial	5.6	2.0	9.3	17.5	4.5	13.1					19.4	32.6	2.0	11.0	21.4	43.4
Transportation: propulsion only	0.1		0.6	1.0	14.5	35.6					15.2	36.6	0.02	0.1	15.2	36.7
Electricity generation utilities	7.1	24.2	3.2	4.1	1.2	0.9	2.4	5.1	0.13	38.1	14.0	72.3	(4.5)	(30.8)		
Totals[a]	13.4	26.2	19.6	41.7	27.1	51.6	2.4	5.1	0.13	38.1	62.4	162.6			52.9	121.1

[a]Miscellaneous included. Entries may not add to total because of rounding off of numbers.
Source: Mineral Facts and Problems, Bureau of Mines Bulletin 650, p. 17, 1970.

transport propulsion energy is expected to increase from 29% of the U.S. Net Energy Input in 1968 to 30% of NEI in 2000, or 36.7 quadrillion BTUs (QBTUs). NEI, as previously explained, is roughly equivalent to total U.S. energy production minus that lost in the generation of electricity by utilities. In the same period the Gross Energy Input into the electricity generating sector is expected to increase from 14 QBTUs to 72.3 QBTUs. Petroleum will continue to supply most of the propulsion energy, with electrical energy contributing only about 0.3% of NEI by 2000. It is estimated that consumption of electricity by the secondary-transport-related activities will rise to 2.3% of NEI; thus, the overall demand for electricity by the transportation sector will remain at the current level of 11% of electricity purchased (Grimmer and Luszczynski, 1971).

The share of the total U.S. petroleum used for propulsion is expected to increase from 54% in 1968 to 69% in the year 2000. These projections are based on the assumption that the present trends will continue in the absence of any significant constraints on fuel supply or changes in propulsion systems. The annual rate of increase in use of transport petroleum is similar to that of the U.S. Gross Energy Input, but is considerably higher than the growth rate of the population. For example, in 1950, 360 gallons of transportation fuel were used per capita. This figure is expected to double by 1980.

The annual amounts for GEI, total petroleum, and transport petroleum are increasing at a rate corresponding to a doubling period of eighteen years, while, by comparison, electricity purchased per year doubles in less than ten years. Doubling of the U.S. population may take more than fifty years (Dept. of Commerce, 1970). It should be noted that during one doubling period of the annual demand for electricity, as an example, the cumulative demand also doubles if the increase is constant. Thus, during the next ten years the cumulative consumption of electricity is expected to be equal to all the electric energy consumed to date.

The demand for transport petroleum is closely linked with the distribution of the various transportation modes.

In passenger transport there is a rapid decline of travel by train. The fraction of travel by busses is also declining, while air travel is increasing at a fast rate with some leveling-off in recent years. In 1968 automobiles accounted for 87% of intercity travel; the annual numbers of passenger miles traveled by automobile between cities have shown a steady increase corresponding to a doubling period of fifteen years.

In cargo transport, railroads moved 41% of intercity cargo in 1968. Air cargo transport has had the highest growth rate over the past twenty years, but by 1968 it only accounted for less than 0.2% of cargo. Pipelines are increasing their share of intercity transport, possibly at the expense of trucks and railroads. Pipeline transport accounted for 21.3% of total intercity cargo tons shipped in 1968, compared with 21.6% for trucks and 15.9% for watercraft. Waterway cargo transport shows steady growth.

In urban areas almost all transportation is provided by motor vehicles. Automobiles are providing over 80% of the total vehicle-miles. The relative role of busses is declining. Since 1960 the increase of annual motor vehicle-miles has been fairly steady, indicating a doubling period of about fifteen years, with an expected doubling in twenty years of passenger-miles per capita. Clearly, whether this increase in transportation and mobility can continue much longer at the current rate will depend, among other things, on the availability of transport energy, which is almost entirely based on petroleum.

The subject of petroleum reserves has received much attention for many years. Yet there appears to be a great deal of uncertainty about the actual amount of recoverable resources. Estimates of the potential U.S. oil reserves, exclusive of Alaska, range from 145 to 532 billion barrels, while figures quoted for Alaska vary from 5 to 75 billion barrels (Bureau of Mines, 1970). The figure of 200 billion barrels, given by M. K. Hubbert (1971) is probably representative of the readily recoverable U.S. crude oil reserves. This figure, when converted to energy units, at 130,000 BTUs per gallon, amounts to 1,090 QBTUs. Deposits in Alaska are approximately one-tenth of that, or 100 QBTUs. The petroleum resources of the world are estimated to be on the order of 10,000 QBTUs (Hubbert, 1971).

Estimates of ultimate petroleum reserves, however, are of less significance than the technological and economic factors which influence discovery of resources and conversion to proved reserves. (Proved reserves are a working inventory of oil that has already been discovered and is recoverable under existing conditions) (Bureau of Mines, 1970). The average U.S. petroleum reserve-production ratio has been declining from its historic levels of twelve to less than ten in 1968. In other words, there was ten times more petroleum in reserve than was produced in 1968, whereas formerly there was twelve times more in reserve than was produced. Further decline is expected in the future.

At the end of 1968, the estimated recoverable reserves from crude oil reservoirs in the U.S. totaled only 208 QBTUs (Bureau of Mines, 1970). By 1985, the annual total petroleum consumption is expected to be about 45 QBTUs per year, with about half of this going into transportation. The cumulative use for petroleum between 1968 and 2000 is expected to exceed 1060 QBTUs (Mills et al., 1971).

Inasmuch as it is difficult to comprehend the very large quantities of energy needed for transportation now and in the future, it might be useful to observe that oil reserves of Alaska, in the neighborhood of

100 QBTUs, could provide only a seven-year supply of transport fuel in 1968, a four-year supply in 1985, and less than a three-year supply by the year 2000, according to projections made by the Bureau of Mines. By 1985 the total annual U.S. petroleum demand thus might be equivalent to using up the energy equivalent of the total potential oil reserves in Alaska every two years. The *cumulative* use for the total U.S. petroleum by the year 2000 might be more than ten times the amount of reserves in Alaska.

From the current data, therefore, it appears that fuel constraints will become important in the development of transportation. Given these conditions, it is necessary to consider various methods for reducing fuel costs per unit of transportation.

REQUIREMENTS OF ELECTRIC TRANSPORT

It is important to know how much energy is really necessary for a given amount of transportation. Electrified transport provides a convenient reference since the overall efficiency and energy losses of electric vehicles can be readily evaluated. Also, it is of some interest to estimate how much electricity would be required for a partial or total electrification of land transportation.

The electric vehicle has attracted much attention, and there are numerous estimates of its performance under various conditions (Netschert, 1970; Agarwal, 1971; Morse, 1967; *Electric Vehicles,* 1967). The figures given by Netschert are probably representative of high-performance electric vehicles. Netschert assumes the efficiency of the electric motor-control system to be 90% and that of the battery to be 80%. Additional losses in delivering energy to the wheels might be of the order of 10%. These figures lead to a net efficiency of 65% for the electric vehicle. That is, 65% of the electrical energy supplied to the evhicle is used in actual propulsion, and 35% is wasted.

To evaluate the overall efficiency of the electric vehicle, it is necessary to take into account the energy lost in the generation and transmission of the electricity that is eventually used to propel the vehicle.

In the case of coal power plants, for example, one must consider the efficiency at each stage in the flow of energy from production of coal to consumption of electricity. Efficiencies of coal production and transportation appear to be 96% and 97% respectively (Ayres and Scarlot, 1952), implying that by the time coal reaches the power plant about 7% of the raw fuel energy has already been expended.

In 1968 the electricity purchased from utilities represented 32% of the Gross Energy Input into this sector (or, in Table 1, the ratio of 4.5 QBTUs of Electricity Sold, to 14 QBTUs for Electricity Generation, Utilities). Combination of all the steps in the production and distribution of electricity gives a system efficiency of about 30%. In other words, less than one-third of the raw fuel energy is converted into available electricity. When we account for the further energy losses in the vehicle propulsion system, the energy-system efficiency of the electric car is only about 20%. The energy flow and losses in the entire system are represented schematically in Figure 2.

Thus, efficiency of the electric car depends directly on the efficiency of the electric power plant. As the generation of electricity and utilization of waste heat from power plants are improved, the system efficiency of the electric vehicle will increase. Utilization of waste heat from power plants located in urban area appears to offer considerable potential for improvement in this regard, so the 20% efficiency for the electric vehicle represents a low estimate.

It is interesting to compare these results with the performance of the gasoline car. Ayres and Scarlot (1952), using pre-1950 figures, estimated the following efficiencies: production of crude petroleum, 96%; refining, 87%; transportation of gasoline, 97%. In other words, about *one-fifth* of the fuel energy is effectively lost by the time gasoline is pumped into the car. Estimates of performance of the vehicle itself give 25% for thermal efficiency of the engine, 70% for its mechanical efficiency, and a 70% efficiency for transmission of power to the wheels (SAE Symposium, 1956). The resulting efficiency of the vehicle is in the range of 12%. Thus, the *energy-system efficiency* of the gasoline car is in the neighborhood of 10%. While these numbers are rather approximate—since the actual performance depends on many factors—it is clear that only about one-tenth of the raw fuel energy is delivered to the wheels of the gasoline car, as compared to one-fifth to the electric car. There appears to be little potential for significant improvement in the overall efficiency of the gasoline car without radical changes in design. Its efficiency is further impaired by restrictions on emissions. By comparison, the electric car offers considerable promise for improvement, especially if waste heat from electricity generation can be at least partially utilized.

In addition to a number of practical difficulties with the electric car which center primarily around the battery design (Morse, 1967; *Electric Vehicles*, 1967), there is the question of added demand for electricity. Currently, most of the transportation energy is derived from petroleum. About 70% of this petroleum is used by motor vehicles. In 1968 there were 101 million motor vehicles registered in the U.S.—83.7 million automobiles, 17 million trucks, and 350,000 busses. The average installed power per vehicle was 60 horsepower, giving a total installed automotive power of 12.6 billion kilowatts. This should be compared with the total installed electric power capacity of 276 million kilowatts. While the fuel energy used by motor vehicles was 10.7 QBTUs (Table

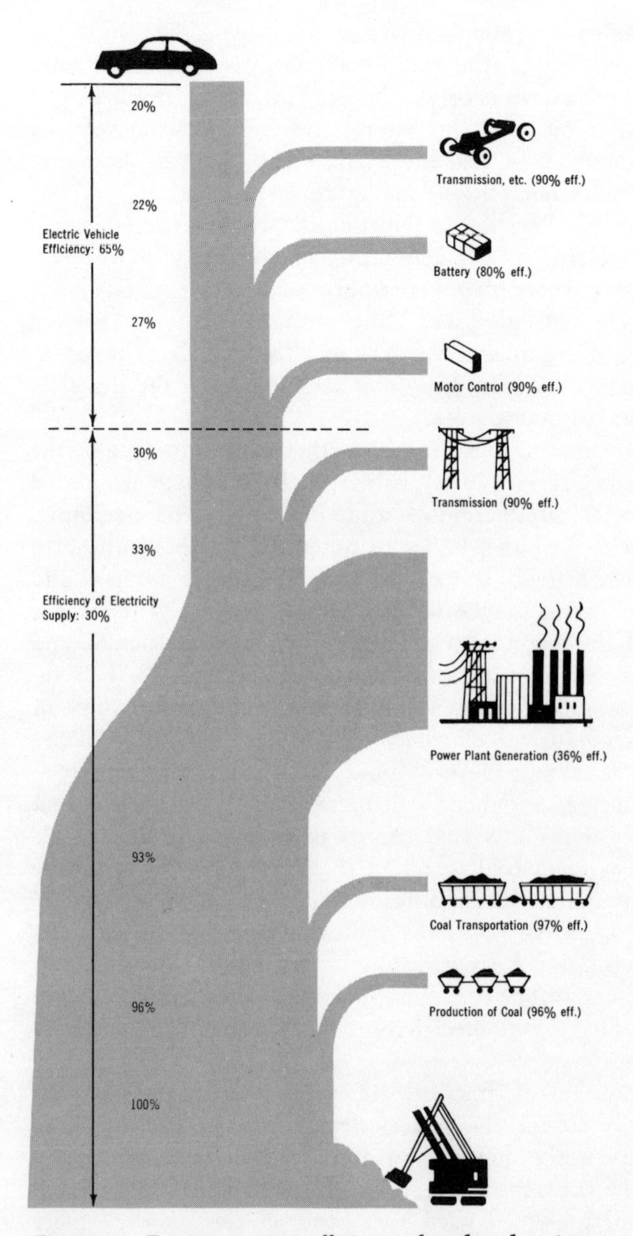

Figure 2 Energy-system efficiency for the electric car

2), the electric utilities expended 14 QBTUs (selling 4.5 QBTUs and wasting 9.5 QBTUs—see Table 1). Thus, though the installed automotive power was 46 times more than that of the electric power stations, the amounts of energy expended by transportation and electric utilities were of the same order. This arises from the fact that only about 3% of the motor vehicle capacity is used, whereas power stations run at an average of 52% of capacity.

To the extent that the amount of energy used by motor vehicles is more than two times greater than the electricity purchased (10.7 QBTUs versus 4.5 QBTUs of electricity sold), it would appear at first glance that electrification of transport cannot be seriously considered without a large expansion of the electric power capacity. A closer examination of this question shows, however, that there would be an added demand of only a fraction of the electricity now purchased.

Using certain broad assumptions it is possible to derive energy requirements for various vehicles (Netschert, 1970; Morse, 1967). Netschert estimates that a 4000-pound electric car traveling at 40 mph would require 1200 BTUs per mile to overcome air resistance, rolling resistance, and road grades; this estimate also includes an acceleration factor and an accessory load. This figure can be compared with the actual energy requirements of a 4000 pound gasoline car (SAE Symposium, 1957; Baumeister, 1970). At 25 mph, it takes about 4 horsepower to drive this car; at 30, 40 and 60 mph the power requirements rise to 7, 11, and 29 horsepower respectively. The power required to drive the car at a steady speed of 40 mph corresponds to 700 BTUs per mile. In urban areas the average speed of motor vehicles is usually well below 25 mph. At this speed the energy requirement is 410 BTUs per mile. Acceleration, road grades, and accessories will tend to raise this figure, though with electric propulsion some of the energy expended on acceleration and grades can in fact be recovered (*SAE Journal*, 1970). Also, unlike the gasoline car, the electric vehicle does not use any energy when stopped, except for accessories; this can be an important factor in high-density urban traffic where a good part of travel time is now spent under idling or near-idling conditions. In fact, in traveling through urban areas, about one-third of the time is spent on idling or deceleration. Thus, the energy requirement of 1200 BTUs per mile used by Netschert appears to be quite adequate and possibly even too high.

Assuming that each of the 83.7 million automobiles in 1968 drove an average of 9500 miles (Dept. of Commerce, 1970), the resultant energy requirement that year was 1 QBTU. It should be noted that the actual fuel used was 7.6 QBTUs, implying an apparent vehicle efficiency of 13%. Similarly, for trucks and busses Netschert derives an energy requirement of 2800 BTUs per mile, or the annual total of 0.56 QBTU for 350,000 busses and 17 million trucks. Thus, the total motor vehicle energy requirement in 1968 was around 1.56 QBTUs. If the vehicles were electrified and if the vehicle efficiency were in the range of 65% the demand for electricity would have been 2.4 QBTUs. If just the urban motor vehicles were electrified, the additional demand for electricity would amount to 1 QBTU, or 22% of electricity purchased in 1968.

Railroads consumed about 0.72 QBTU of energy in 1968. At most, 30% of this energy, or 0.22 QBTU, was delivered to the wheels. This means that electrification of half of all the railroads would have added 0.12 QBTU to the demand for electricity, allowing for 10% losses in the transmission of electric power to the wheels.

Thus, the additional electricity that would be required to electrify urban motor vehicles and half of the railroads would amount to 1.12 QBTUs, or about one-fourth of the electricity purchased in 1968. Although this is a large increase in demand for electricity, to the extent that much of the increased demand would be for charging batteries or other energy-storage devices during off-peak hours, the existing electric capacity might be adequate in many localities. This additional demand would also tend to smooth out the load on power plants, permitting more economic operation and resulting in lower costs per unit of electricity.

In conclusion, since the actual energy requirements for propulsion on land are relatively low, a substantial electrification of transportation appears possible without a massive expansion of the electricity-generating capacity. Electrification of propulsion is expected to result in lower fuel costs per unit of transportation.

FUTURE DEVELOPMENT OF TRANSPORTATION

The energy used in transportation places a heavy demand on the known fuel resources. Although the potential U.S. reserves might be adequate for several decades, continuation of the current increase in use of transportation, which derives most of its energy from petroleum, is not possible without substantial additions to the known U.S. petroleum reserves.

Given the increase of total and per capita transportation as well as the known constraints on fuel resources, future development of transportation will require direction and comprehensive planning. In addition to other factors it will clearly be necessary to consider the efficiencies of the different modes of transportation. The present lack of regard (especially evident in the category of passenger transport) for this aspect cannot continue indefinitely, and the current trend to lower efficiency will have to be reversed. In this context, it is apparent that existing or new transportation modes which do not offer any significant reduction of energy used per unit of transportation cannot provide satisfactory solutions.

Considerable savings of energy are possible in the transportation sector through enhancement of high-efficiency modes such as busses and trains in passenger transport, and pipelines, waterways, and railroads in cargo transport. As a rule, higher efficiency is obtained with those modes which offer higher ratios of payload to gross weight.

Energy requirements for land transportation can be reduced by electrification of propulsion, especially in areas with high density traffic.

Since the actual energy requirement of electric propulsion is small when compared with the energy used by current vehicles, most of the urban motor vehicle transport and half of all the railroad transport could in fact be electrified with an increase of about one-fourth of the annual electricity supplied by utilities; this electrification could be accomplished without a massive expansion of the installed electricity-generating capacity, which is now only partially utilized.

In the final analysis, there is little doubt that energy requirements of transportation demand careful attention, but the choice of particular methods employed to optimize energy use in transportation must depend also on social, economic, and environmental factors.

REFERENCES

Agarwal, P. D., 1971. Society of Automotive Engineers (SAE) Meeting, Detroit, Mich. Jan.

American Automobile Association, 1970. *Automotive facts and figures.*

Ayres, E., and C. A. Scarlot, 1952. *Energy sources—The wealth of the world.* New York: McGraw-Hill.

Baumeister, T., ed, 1970. *Standard handbook for mechanical engineers.* New York: McGraw-Hill.

Bureau of Mines, 1968. *Minerals yearbook.*

———, 1970. *Mineral facts and problems.* Bulletin 650.

Department of Commerce, 1970. *Statistical abstract of the U.S.*

Electric Vehicles, etc., 1967. Hearings, Committee on Public Works, U.S. Senate, Mar.-Apr.

Grimmer, D. P., and K. Luszczynski, 1971. Energy use—A look at transporation. St. Louis: Washington Univ., unpublished.

Hubbert, M. K., 1971. *Sci. Am.* Sept.: 61.

Mills, A., et al., 1971. *Environ. Sci. Technol.* 5 (Jan.): 30.

Morse, R. S., 1967. *The automobile and air pollution.* Dept. of Commerce.

Netschert, B. C., 1970. *Bull. At. Sci.* May: 29.

Rice, R. A., 1970. ASME, 70 WA/ENER-8, Nov.

SAE Journal, 1970. p. 42, Aug.

Society of Automotive Engineers (SAE) Symposium, June 1956. Where does all the power go?. *SAE Trans.* 65 (1957): 713.

RICHARD G. STEIN

Energy Use in Architecture and Building[1]

In 1970 the U.S. had a Gross National Product of just under $1000 billion. About $100 billion or 10% of this was provided by the construction industry (building construction, that is, not including roads and highways) (*Sci. Am.*, 1970). In 1970 in the U.S., a total of 69×10^{15} BTUs of energy were consumed. This included about 1.7×10^{12} kwh of electricity, which can be expressed as 5.8×10^{15} BTU's. Of the total electricity produced, 128,000 million kwh was used by the construction industry, either directly in construction or indirectly through production of the materials used in construction. This is 7.5% of all electricity produced and 22% of electricity used by industry (Fig. 1). In addition, the built-in standards for lighting, heating, cooling and other electrical energy uses required to operate the buildings represent just under 50% of all electrical usage and therefore about 12% of total energy usage (Fig. 1).

In addition to this sizable amount of electric power consumed directly by building construction, architecture, through its product, the man-made environment, has a greater influence on energy use than any other major component of our GNP except transportation and the military. Thus it is important to examine the full interrelationship between architecture and energy, in search of energy savings that can be achieved in new and existing buildings and the conditions under which these energy savings can be made. To do this it is necessary to determine how and where the energy use commitments are made; how extensive they are and what can be done to modify significantly the

[1]Important help and data in preparing this paper were furnished by Diane Serber, AIA; Peter Flack, P.E.; and Michael Corr among many others.

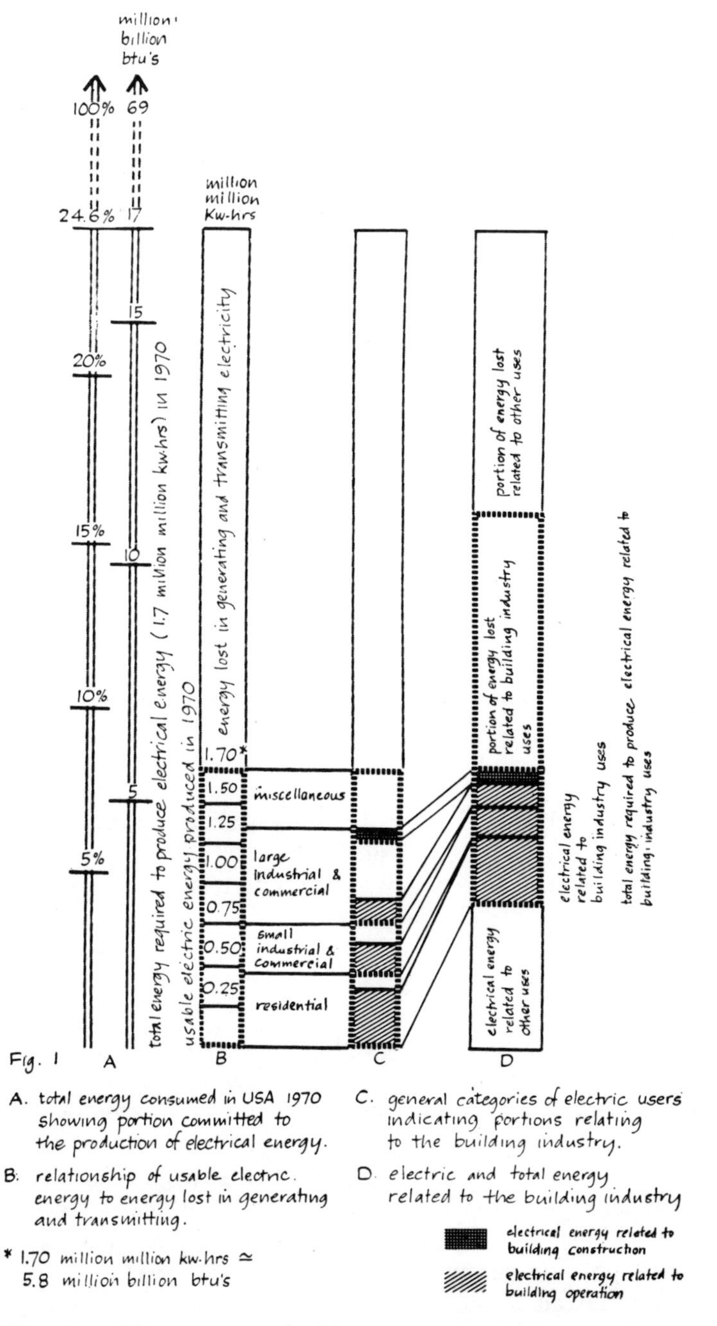

Figure 1 Electric use in buildings as part of total energy use

energy (particularly electrical) required for constructing, maintaining and demolishing buildings.

There are essentially four broad areas of energy use in the life of a building. At the outset, energy is required for the production of the materials that will be used in construction. Next, energy is used to assemble the materials into the finished inhabitable product. Third, maintaining and operating the building consumes energy. And finally, the demolition of the building is an energy consuming process. As each of these areas is influenced by the others, and all are dependent upon the design of the building, energy consumption by the building throughout its life is to some extent the responsibility of the architect.

THE DESIGN OF BUILDINGS—MATERIALS AND CONSTRUCTION

From the very outset of the architectural process energy is used wastefully. Our basic structural sciences are in reality more recordings of practical experience with large safety factors than they are scientific analyses of how material is used. Steel for example, is designed for the convenience of the rolling process. A steel beam is designed to resist failure at the critical point of its span. The configuration of the beam, the cross section, is generalized for reasons of manufacture and is carried through unchanged from end to end. Only in large structures, or in those for which weight is an important factor—trusses, ships, planes, bridges and such—is there an effort to place material where it approximates the efficient form for the structural job required. Changes in fabrication methods and greater repetition of structural components can reduce the amount of steel needed, hence the amount of energy used to produce it, to perform largely the same job that is now being called for.

In addition, all structural design, an empirical process, bears the weight of a pyramiding set of safety factors. Let us look at a simple concrete beam computation as an example. In figuring the loading, live-loads (that is, loads other than the weight of the structure) are assumed to be simultaneously applied over all rooms, corridors, lobbies and stairs. A 750 square foot classroom for 30 pupils is computed to withstand a load of 40 pounds on every square foot, or a total of 30,000 pounds in addition to the weight of the structure (National Board of Fire Underwriters, 1955). Thirty large children and a teacher might weigh 5000 pounds. Thirty-one desks and chairs might add another 3000 pounds. Adding another 1500 pounds for friends, books, paraphernalia, and miscellany would bring it to 9500 pounds, less than one-third of the figure used in the computation. In addition, the values given for the concrete have a 300% safety factor and for steel a 50%

factor. (Concrete with a rated strength of 300 psi has an ultimate strength of 3750 psi. In structural computations, however, the design strength used for this concrete is 1350 psi, only 36% of its ultimate strength. Steel is designed at 60% of its yield point, i.e., 36,000 psi steel is designed for a maximum stress of 21,600 psi.) Furthermore, the structural designer will select available steel for reinforcement and overall dimensions for his beam cross-section at the first size above the size required by the computation, adding, on the average, another 5%. And then of course, the concrete works in a much more complex manner than is taken into account in the individual and isolated design of each member. The concrete, for example, below the neutral axis is not given any structural consideration. Even discounting the exaggerated loadings that are contained in most liveload tables which become incredibly high when one compares them with the maximum loads that the building actually sustains, it is quite obvious that reducing the safety factors would permit concrete to be designed with less than half the material now used. If you discuss this with any structural engineer, he will confirm this under several provisos. First, that building codes be rewritten. Second, that there be enough in the budget to pay for the labor that is necessary to build formwork carefully and to place steel carefully and to mix and place concrete carefully. In addition to the savings in the individual members, there is a further cumulative saving since the weight of the building itself is substantially reduced. The size of the footings and foundations can be considerably reduced, with further savings in material, reflecting both the more realistic structural analysis and the reduced loadings that the foundations and footings are designed to support.

We have computed that in cement production alone this could result in energy savings of about 20 thousand million kwh a year. To make a comparison, an electricity usage budget for an average family is in the range of 7000 kwh per year. Thus, this savings alone would provide the electric power for 3 million families.

Another large area of investigation is concerned with the different energy consumption required by different materials. Synthetics and plastics in general require more energy than the natural materials they replace because the process of making the plastics and synthetics requires the applicaton of large amounts of energy to the basic materials, petrochemicals, in order to break them down and rearrange their molecular structure into filaments and powders. In the materials that the plastics replace, these energy infusions have been the result of solar energy, through the photo-synthetic process.

Other materials, such as aluminum require a great deal of energy in their manufacture since the aluminum refining process has a major

Table 1 Electrical Use in Building Construction

	Total Sales to Building Construction Industry (Excluding Highways) in GNP (million $)	Total Cost/ $1,000 for All Electric Energy	Cost/ $1,000 for Electric Energy, Applied Directly	Total Cost. All Electric Energy (million $)	Cost. Electric Energy Applied Directly (million $)
Electric lighting and wiring equipment	1,786	17.60	4.61	31.5	8.25
Heating, plumbing, structural metal products	9,557.6	21.10	4.33	201	46.2
Other fabricated metal products	1,524.4	20.07	6.39	31.6	9.72
Primary copper manufacturing	2,007.6	28.9	6.55	58.2	13.13
Primary aluminum manufacturing	78.2	61.4	30.63	4.8	2.4
Primary iron and steel manufacturing	2,950.6	28	13.90	82.5	41
Stone and clay products	8,785.3	27.20	14.82	239	130
Lumber and wood products except containers	5,872	16	6.63	94	39
Petroleum refining and related industries	1,303.9	16.3	5.38	21.3	7.02
Electric utilities	214.6	107.3	56.02	23	12
Gas utilities	42	9.6	0.94	0.4	0.04
Motor freight transportation and warehousing	1,719.7	8.5	3.37	14.6	5.79
Wholesale and retail trade	8,530.2	17.1	12.34	145	105
Business services	4,050.1	15.5	7.83	62.8	31.6

This represents electric energy use in 88.5% of the construction industry's share of the GNP (excluding highways); 49.3% represents materials; 39.2% represents value added. Extrapolating for 100% produces a figure of 128,000 million kwh. Total U.S. production of electric energy in 1969 (according to Edison Electric Institute) was 1,556,996 kwh.

Cost. Electric Energy Applied Indirectly (million $)	Rate. Electric Energy This Industry (¢/kw-hr)	Average Rate. Electric Energy. All Industries (¢/kw-hr)	Electric Energy Directly Used (million kw-hrs)	Electric Energy Indirectly Used (million kw-hrs)	Total Electric Energy (million kwh)
23.25	1.05	0.875	790	2,660	3,450
154.8	0.78	0.875	5,920	17,600	23,520
21.88	1.43	0.875	680	2,500	3,180
44.9	0.74	0.875	1,780	5,130	6,910
2.4	0.32	0.875	720	280	1,000
41.5	0.78	0.875	5,250	4,750	10,000
109	1.04	0.875	12,500	12,500	25,000
55	1.23	0.875	3,160	6,290	9,450
14.28	0.74	0.875	950	1,630	2,580
11	0.87	0.875	1,380	1,260	2,640
0.36	0.12	0.875	30	40	70
8.81	0.875	0.875	660	1,010	1,670
40	0.875	0.875	12,000	4,570	16,570
31.2	0.875	0.875	3,620	3,570	7,190
					113,230 million kwh

Sources: Scientific American, "The Input/Output Structure of the United States Economy," 1970; U.S. Census of Manufacturers, 1967.

electrolytic component. Aluminum requires about five times the energy per pound that is required in the manufacture of steel. Obviously, the two are not universally interchangeable nor are they interchangeable on a pound for pound basis. However, as an example, the 4 million pounds of aluminum that have been listed (*Design Newsletter*, 1971) as being required for the skin of a new office building in Chicago, could be replaced with about 5.75 million pounds of stainless steel which would have approximately the same structural characteristics and the same weathering characteristics (Aluminum Association, 1970). The weight is somewhat greater but the gauge would be a little less. In energy terms however, the aluminum would require 2.1 million kwh to process and assemble, about three times the 0.77 million kwh necessary for the stainless steel. The difference would be about 1.3 million kwh on this building alone. As another example, a damaged section of a heavy aluminum pipe guard rail on a road interchange in New York City has been replaced with a galvanized steel assembly (St. Joe Minerals Corp., 1971). While energy conservation was probably not the motivating reason for the change, the material in the aluminum rail required 2.2 million BTUs per 20 foot section against the material in the steel rail's 0.64 million BTUs for the same length.

As the importance of energy saving in building material use becomes more widely accepted it will also have a profound effect on building aesthetics and evaluative standards of architectural critics. There will be a renewed enjoyment of the taut, the tense, the spare and a falling off of the rhetorical, the over-convoluted, the use of complication of form for the surface interest it generates. The earlier attitude produced the Crystal Palace, Brooklyn Bridge, Model A Ford—the latter the Yale School of Architecture, the fish-tail Cadillac, and the electric carving knife. As this attitude takes over I believe the architectural profession will be able to satisfy performance needs with no diminution in standards, with a saving of 10% in materials used and, therefore, in the energy required to produce them.

So far, we have been talking about using materials efficiently and intensively within the constraints that are set up by building practice, codes and by economic comparisons. As we look through the publications however, we cannot fail to be impressed with the excessive amount of material used purely because of architectural decisions. These go far beyond the material usages that would result from careful design consideration, even under present limitations. It is hard to know how to quantify this arbitrary overuse of materials but any architect, I am sure, could cite several examples of materials that have been used unnecessarily, redundantly or excessively.

MAINTENANCE AND OPERATION OF THE BUILDING

Much of the energy demand that goes into the operation and maintenance of a building is predetermined by the building design. Chief among these are lighting, heating, and cooling. Variations in design can, without impairing the operation or quality of the building, significantly affect the amount of energy that is needed by the building throughout its useful life span.

Design of Lighting

Lighting deserves especially careful study. It is at once the most visible use of electrical energy and from a quantitative point of view, it is responsible for 24% (according to Cory N. Crysler of General Electric) of all the electricity sold in the United States. In New York City, because of the large amount of commercial space, the percentage of the electricity produced that goes into lighting was cited by Con Edison as being as high as 65%. Looking at these two percentages, it becomes apparent that any savings due to elimination of unnecessary lighting can do a great deal to relieve the critical electrical energy shortage.

Lighting levels, in general, are provided above the minimum recommended lighting standards of the Illuminating Engineering Society. These standards have been translated into various school codes, state codes and municipal codes, and beyond this, they are the handbook norms that electrical engineers use in the absence of any other data in order to have some documented basis for the light levels they provide. Over the last 15 years, recommended IES light levels have more than tripled. As an example of the increase in levels, New York City's Board of Education's Manual of School Planning called for 20 foot candles in classrooms in 1952. It was raised to 30 in 1957 and to 60 in 1971. Libraries went from 20 in '52 to 40 in '57, to 50 in '59 and to 70 in '71. Shops now require 75 and drafting rooms 100.

The rationale for the increases is contained in a 1959 report published by the IES and based on a study conducted by H. R. Blackwell. The Blackwell report contended that efficiency in performing visual tasks is in direct proportion to the foot candle level of diffuse undifferentiated light achieved in the space.

It is now an IES axiom that contrasts in light levels are bad and a comfortable light environment depends on keeping intensities within a 3 to 1 ratio of either side of the required light level. There has been no physiological or psychological verification of this assumption. In fact, physiologically our eyes adapt well to the contrast of 1500 foot candles of the sun on the snow and 20 foot candles in the shadows of a pine tree's branches. On the other hand a non-contrasting environment, carried to its ultimate form of a white shadowless environment, has resulted, in laboratory tests, in short term aberrations, hallucinations, and extreme mental discomfort. More to the point, it is now a textbook axiom that a monotonous environment can lead to boredom and loss of productivity (Hebb, 1972).

Since contrasts in light levels within the field of vision were considered detrimental, total area illumination to the levels set for particular tasks became the general practice in the lighting field. This tendency is becoming still more dominant with the publication of a newer concept by the IES, called the Equivalent Sphere Illumination or ESI, in which the lighting will attempt to stimulate a sphere of illumination with the task to be lit at its focus, requiring considerably higher inputs than even those now recommended (Fischer, 1971). In fact, a body of research exists in opposition to the design philosophy developed by the lighting industry. Tests conducted by Tinker at the University of Minnesota in 1938, and by Butler and Rusumore at San Jose College in 1968, conclude that a light level between 3 and 10 foot candles is adequate for efficient reading and that higher light levels do not increase efficiency and may increase fatigue. Tinker states, "The experiments of Luckiesh and Moss have led them to recommend what seems to be excessively high light intensities for visual tasks . . . 20–50 fc for ordinary reading . . . Reexamination of their data suggests that their conclusions are not justified." Tinker concludes his paper: "10–15 fc should provide hygienic conditions when one's eyes are normal and print is legible. Fine discriminations require 20–25 fc for adequate vision." His observations seem to be borne out by the recent lighting renovation in St. Patrick's Cathedral in New York City (Steiner, n.d.). In order to allow everyone, including the elderly, to participate in the Mass, the lighting level was raised from 4 fc to 17 fc. Thus the difference between 17 fc and the levels recommended by the lighting industry are not necessary for the performance of most visual tasks and are provided only for the ambience or impression they create. In line with this attitude, a lighting fixture advertisement in a recent issue of a construction industry magazine recommends "minimum lighting levels of 90 fc throughout the building" (GTE Sylvania, 1971) for new office buildings, not as a performance requirement, but for its cosmetic effect. The prevailing attitude, too, that a high light level should be generalized throughout an area rather than being directed to the work area, as it is in an operating room, again substitutes large scale energy input for specificity.

Further energy savings can be seen in a reinvestigation of the light producing component itself, the lamp. The fluorescent lamp delivers three times the lumens per watt of a standard filament bulb. However, even though it is inherently an effective light fixture because of the large illuminated surface area over which the light source is distributed, it is usually placed in a diffusing container, the luminaire, or fixture. If a plastic louver with a 45° cutoff is added to the bare tube, it reduces its output to 76%. A translucent plastic cover reduces the level to 65%. If the tube is mounted in a suspended semi-indirect luminaire with plastic sides and bottom, it will only deliver 25% of its bare tube potential (Westinghouse Electric Corp., 1960). As an example, in the 1950s, the New York City subway system installed continuous flourescents with cast glass fixtures above the platform edge in most stations. Some ten years later all cast glass covers were removed and discarded and light levels increased noticeably with no increase in energy input.

While some shielding may be desirable to reduce glare and while occupants of a space may be more comfortable at light levels above the minimum physiologically necessary for the task, the present recommendations are absurd.

Careful study of the placement of light sources reveals further potential reductions in installed capacity. For example, Educational Facilities Laboratory sponsored a study that demonstrated that by departing from the traditional continuous strips of lighting over an entire classroom ceiling and placing the luminaires at the perimeter, a 40 fc level was as effective as 60 fc typically provided (Sampson, 1970). A large amount of the electrical energy in lighting is unnecessarily used to light the outer strip of a building, a section that can be adequately lighted by natural light. Most of this energy could be saved by circuiting the perimeter separately and by having the lights turned on and off by a light sensitive switch. Where separate switching has been provided, experience indicates that we Americans are not sufficiently motivated to do it manually.

To summarize the above, it appears—conservatively—that adequate lighting could be installed in institutions, commercial buildings, schools and so forth with less than 50% of present light loads. This does not take into account additional savings that could be realized by having more selective switching arrangements, by eliminating electrical light usage when adequate daylighting exists, nor does it take into account the air-conditioning savings that would result if the plants did not have to remove the heat that the unnecessary lights brought into spaces. Without considering the savings in direct electrical usage at the moment, there is also the savings resulting from fewer electric fixtures, smaller wiring loads and reduced sizes of switch gear. Of some 5400 million kwh that are used in the building construction industry for the electric lighting and wiring components of buildings, at least 25% or 1350 million kwh can be saved.

ELECTRIC HEATING

Next, let us consider the use of electricity for heating. In the last five or six years there has been an accelerated drive by the utility companies to increase the number of electric heating installations. The campaign has been successful. The residential consumer, while still not the largest user of energy, is the group that is growing most rapidly. In 1960 it represented less than 29% of the market buying electricity. Its share of the market grew slowly to 30% by 1968 and jumped to 32.2% by 1970.

The major factor accelerating the growth is the great increase in electric heating. In 1970, with a 7.5% saturation of the residential market, electric home heating throughout the U.S. used about 66,000 million kwh of electricity—almost 4% of all electrical energy produced and sold (EEI, 1970).[2] If the ratio of capacity to consumption of Con Edison generators is used for comparison, a total of about 16 million kw capacity would be required to power home heating; this is equivalent to twice the total capacity of all Con Edison's generators.

According to *Electric World*, the preference for electric heating has increased from 22% to 36% during the last six years. The average consumption of electricity for heating a house is about 15,000 kwh a year.[2] If the rate of 2 million housing starts a year were just maintained, and if the percentage using electric heat did not continue to rise, the commitment to electric energy would be 10,800 million kwh additional every year, for home heating alone, or 108 thousand million kwh in a decade. This figure, totalling 11,300,000 units is an alarming but conservative projection. In fact, the electrical industry projects that the number of electrically heated residential units will be 19 million by 1980 and 25 million by 1985, requiring 375,000 million kwh per year or 85% of residential electricity sales in 1970 (*Electric World*, 1970).

Projections for electric heating for non-residential construction are equally impressive. Of all new non-residential construction reported in 1969, 23% was electrically heated, a total of almost 40,000 buildings, of which 5400 were new office buildings (Thompson, 1970).

[2]4.7 million housing units have electric heat, using 15% of the total national residential usage of electricity. (EPA, N.Y., 1971, p. 179). Households in U.S. in 1970 totalled 62,874,000 (U.S. *Statistical Abstract*, p. 36). Residential electric usage was 32% of 1,394,750 = 440,000 million kwh (EEI *Statistical Yearbook*, 1970, p. 31, 32). 15% of 442,000 = 66,500 million kwh. 62.874 million + 4.7 million = 7.5%.

The use of electrical heating has been made possible largely because of favored electrical rates to all-electric consumers. Rates as much as 50% below those charged to similar but not all-electric users have made electric heating only slightly more expensive than heating by competitive fossil fuels. For example, Con Edison's rate structure for residential users without electric heating is 2.45¢ per kwh over 360 kwh. For users with electric heating, its rate over 360 kwh is 1.90¢ per kwh from May 15–September 15 and 1.40¢ per kwh from September 15–May 15. Below 360 kwh, the rate is the same for both. In general, electric utility companies have insisted on substantially high insulation standards before agreeing to install electric heat, a commendable demand in itself, but one that could reduce the heat load for the competitive fuels as well and the size of heating plants if standards were applied across the board.

Electric heating is inherently inefficient. Fossil fuels such as coal, natural gas, or oil are and will continue to be the prime fuel source for producing electricity for some decades to come. The conversion of fuel energy to heat to produce steam to produce electricity which is then transmitted, with line losses, to an ultimate destination where the energy is converted back into heat, uses the original fuel at less than half the efficiency with which it could be used if converted to heat at its point of use (Gas or oil heating is 70 to 80% efficient, ASHRAE, 1967). In addition, even if there were no inefficiency, the 6% of total electric production used for heating could be allocated for other purposes that can only be satisfied by electric energy.

Only when it appears that the utility companies cannot satisfy the demand they have stimulated will there be a diminution in the campaign for more electric heating. Con Edison has embarked on a Save-a-Watt campaign, and an examiner for the New York State Public Service Commission, Edward L. Block, has been speculating on whether the rate structure should be revised so that large users pay a higher unit cost. To quote the *New York Times* of November 28, 1971:

"When temporary electric rates were approved last June, the Public Service Commission ordered changes in the rate structure, reducing the discounts previously provided to higher volume users.

This was an unusual move aimed at curbing use of electricity during present power shortages. Mr. Block recommended still further reductions in such discounts."

Electric heating is popular with speculative home builders because it reduces the initial cost of the house. No space need be provided for boiler or furnace, nor is the heating equipment required. No chimney or storage tank is necessary. Similarly, builders of apartment and commercial buildings where low construction costs take precedence over long-range disadvantages are attracted to electric heat. This is particularly true where apartments are expected to be co-opted or sold on completion, or where rent structures can be based on having the tenant pay heating costs directly to the utility company, based on individual metering. Even in public housing in New York, a major project is contemplating using electric heat in order to keep visible first costs down and spread the scarce housing building dollars further. The developers of Welfare Island (the project is intended for a population of approximately 20,000 people) are seriously considering the use of electric heating based on economic studies. Gas and steam, as well as electricity are available on the island.

Electricity is advertised as a clean fuel. In reality the pollution is merely removed from the point of use. This problem is graphically apparent in the controversial Four Corners project where the neighbors of the generators at the Utah, Colorado, Nevada, New Mexico line are not the beneficiaries of the power generated, which will be transmitted to Tucson and Southern California, and are actively and vocally opposed to the plant.

ELECTRIC COOLING

Cooling, that is, refrigeration or air conditioning, raises questions similar to those associated with heating. Both have been growing rapidly and use a great deal of electricity.

Cooling, which is responsible today for the summer peak usage of electric energy, can be analyzed in a number of ways. Some are being looked into, others may merit further study. The question of the efficiency of units is the subject of considerable investigation at the moment, and, contrary to the general experience in most fields, there is not only no advantage in oversized individual units, there is actually a collective disadvantage. Air-conditioning units that are oversized tend to satisfy the cooling requirements more rapidly than those designed very close to proper capacity for their cooling job. As a result they tend to cycle much more frequently. When you have a situation as you do in New York where almost a quarter of a million new units are purchased annually, the additional draw that this more frequent cycling requires becomes a significant additional use of electricity. Further, the difference in the delivery of cooling per kilowatt of energy going into the units themselves varies by almost 2 to 1 from unit to unit. Since this is information that is generally not known to purchasers of equipment and since first costs often take precedence over the actual operation and maintenance cost of the unit, there is a tendency to purchase less efficient air-conditioning units that over the long run consume considerably more electricity than they need. In central units, there are alternatives to the electrically driven compressors. These include everything from diesel compressors to lithium bromide absorption units. Even when there is a trade-off in efficiency between the two there is a great advantage in not using an electrical source for cooling since any cooling that is done without adding to the peak demand will not require additional electrical generating capacity.

THE ENERGY DEMANDS OF HIGH RISE BUILDINGS

Since our central cities are highly dependent on electrical energy, we looked into the question of how and where energy is used and committed. In New York City the World Trade Center has an electrical demand of 80,000 kilowatts, more than Schenectady, a city of 100,000, requires (EPA, 1971, p. 168). The extent and implications of the problem are indicated by the amount of new office space built in New York City in the last twenty years—67 million square feet in 195 major office buildings—and the amount projected for the next ten years—an additional 68 million square feet (N.Y. City Planning Comm., p. 31).

With the participation of Peter Flack, a mechanical engineer who adds conscience to competence, we studied a prototypical million-square-foot, 50-story office building to see the effects of certain alternate methods of design. The conclusions are very interesting. Using the design standards that are most widely used for this type of building, and assuming electrical heating and refrigeration, there is an annual consumption of 36.8 million kwh. Of this, 15.8 is required for heating and 3.78 for refrigeration (Computations by Flack and Kurtz, Consulting Engineers, available upon request).

Heating can be supplied by steam with a saving in resource energy similar to the savings in the residential field. Moreover, at least part of the heat may be salvaged from the waste heat in the electrical generating process or from the heat produced in garbage incineration. Cooling can be handled from a central plant rather than with individual air conditioning units; steam-operated lithium bromide absorption units can be used more economically than electric-driven compressor units. Certainly for all cooling tasks of the magnitude of the ones in our study, the use of these units has the double advantage of an overall saving in energy and a curtailing of new demand at the time of peak electrical usage.

If the remaining basic electrical uses are considered as 100%, they are divided as follows: lighting, 54%; advertising and display lighting, 7.3%; elevatoring, 10.9%; fans and air handling equipment, 9.6%; pumps and motors, 4.6%; miscellaneous, 13.6% (Fig. 2).

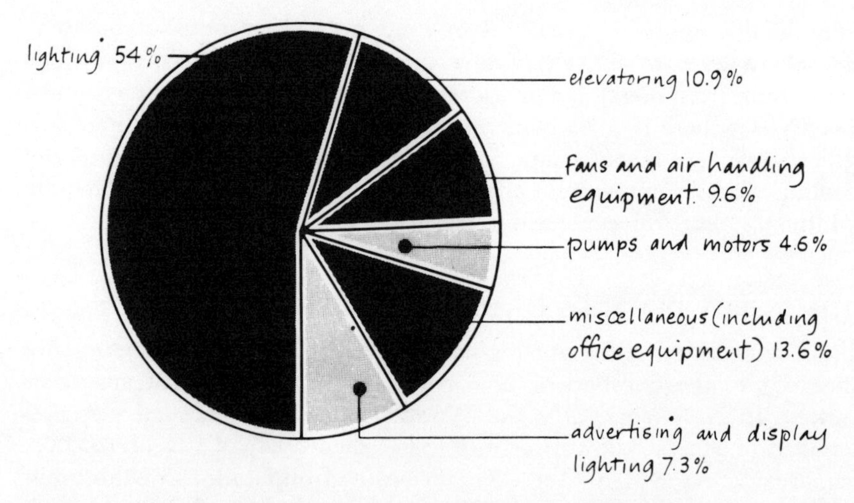

Figure 2 Electric usage in percent for high rise office buildings (with heating and cooling by non-electric means)

If artificial light is not used on a ten foot perimeter strip when there is adequate daylight, the electricity used drops from 9.88 million kwh to 6.79, a saving a 29% on this item. In addition, the refrigeration load is reduced by about 10%, since it is not necessary to remove the heat introduced by the lights. Also there is the saving that would be realized by reducing required light levels as previously described. The lower light levels would reduce the 6.79 million kwh to 3.5 million, 36% of the original load. This would also result in a further saving in refrigeration requirements. Fans to introduce fresh air and remove stale air use 1.75 million kwh. Air movement is based on an empirically established three air changes an hour. Of this 75% can be recirculated and 25% is required to be outside air. In reality, in the winter, building operation personnel will often eliminate outside air on cold days to reduce heating costs. I know of no factual studies that relate health or productivity to air changes, nor is there any consideration given to the quality of the outside air introduced. With operable windows, particularly if there has been some research on the patterns of air movement on the face of these tall buildings, it would be possible to substantially reduce this consumption. The building is occupied for 3100 hours annually. Five hundred of these hours are in the temperature range in which outside air could be introduced with neither heating nor cooling. This in itself would result in a 19% reduction in energy for air handling, a saving of 0.33 million kwh.

Advertising lighting, with 1.31 million kwh allocated to it, merits its own study, for its psychological impact and as a cultural manifestation. Under certain stresses, the whole item can be eliminated, as was demonstrated during WW II, and, inadvertantly, in the great blackout of 1965. When lighting is used for competitive advertising, the tendency is always to raise the ante, until, as in Las Vegas, the amount of lighting is limited only by the amount of surface for attaching the neon tubes and the available techniques for producing the light. As an example, at the other extreme, at Christmas time on Park Avenue in New York a decorative technique has been developed, using a myriad of low intensity bulbs that are strung through branches on the trees. The effect achieved with a modest expenditure of energy is quite captivating. The only discordant note is supplied by a bank building that goes outside the vocabulary and strings garlands of lights with a much higher intensity on its facade, an act that destroys some of the quiet effectiveness of the surrounding displays. The latent saving in the use of light for advertising is enough to warrant an exploration of its ramifications.

The question of savings in elevatoring raises more complicated problems. Lower buildings that do not require vertical transportation have greater surface area with the attendant greater heating and cooling loads. These can more than offset the saving gained by eliminating elevator use. Moreover, lower buildings because of the attendant lower density of population and services, cannot be as well served by mass transportation and may require the extensive use of private automobiles. On the other hand the super highrise buildings are more a product of high land values than of their own inner logic. To give an oversimplified example: in creating a 100-story building, you cannot simply place one 50-story building on top of another. The great bank of elevators that serve the top 50 must be brought down inside the bottom building and along side its elevators. Similarly all its services and pipe systems take up large chunks of expensive space, merely to get where they are needed. Whereas a 50-story building has a roof area with a heat loss on 2% of its floor area, a similar floor area in a three-story building requires a roof area of 33%. The difference between a 50- and 100-story building, however, is only the difference between 1 and 2%.

To further complicate the analysis, there are many other energy-consumptive events that take place when large buildings are grouped in urban centers. The supporting arteries and systems extend farther and farther from the point of use. Water systems, waste disposal, electricity generation, gas lines, sewage systems and deliveries of goods and people extend out and traverse intensively used areas, often bypassing the service of these areas as the upper elevators bypassed the lower. By breaking apart the organism at a certain density, it is possible to create more balanced systems in which the increase in efficiency resulting from the reclaiming of the heat from electric generation for heating and cooling offsets the lesser efficiency of operating more and smaller plants. Since it is heating and cooling that operate to the detriment of the lower buildings versus the 50-story, million-square-foot prototype, such a totally planned scheme would, in its overall context, offer additional energy savings.

To summarize, including all electric usage and using present building standards, 36.8 million kwh are required annually per million square feet. Converted to BTUs, the figure is 129,500 million BTUs. The original energy required to produce this in electricity is 424,000 million BTUs. In contrast, with savings in lighting, cooling and with the other modifications figured, a reduced electric consumption of 13 million kwh is projected, plus 56,000 million BTUs for satisfying heating and cooling needs. The electricity can be converted into 44,400 million BTUs, requiring 145,000 million BTUs of original energy—a total of 201,000 million BTUs—less than half the original 424,000. Extrapolated to satisfy the operating demands of the projected 68 million square feet of New York City office space during the next decade, and not considering transportation, utilities, related residences, governmental buildings and such, 1500 million kwh per year will be required, or, if electricity is used for heating and refrigeration, 2500 million kwh, twice the present capacity of the Indian Point Nuclear Generating Plant. If the projected savings are realized, the electrical consumption will drop to 885 million kwh per year.

DEMOLITION

There is another area of possible savings that ties back into national attitudes and decisions made on other than an energy basis. That has to do with demolition. When we are in a period of unfilled needs nationally in regard to housing, schools, hospitals, cultural facilities, and other buildings, how does one evaluate the decision to replace a building which has a useful anticipated life with one serving the same or another purpose when the decision requires an energy commitment to demolish and dispose of the one building, an energy commitment to build its replacement, and in some cases, an energy commitment for operating and maintaining the new space that is higher than for maintaining the similar-purpose old space.

The Characteristic Energy Curve for Buildings

There is a characteristic curve in all aspects of energy use in the life of a building (Fig. 3). The materials used in building and the construction process can be averaged as a straight line through the construction

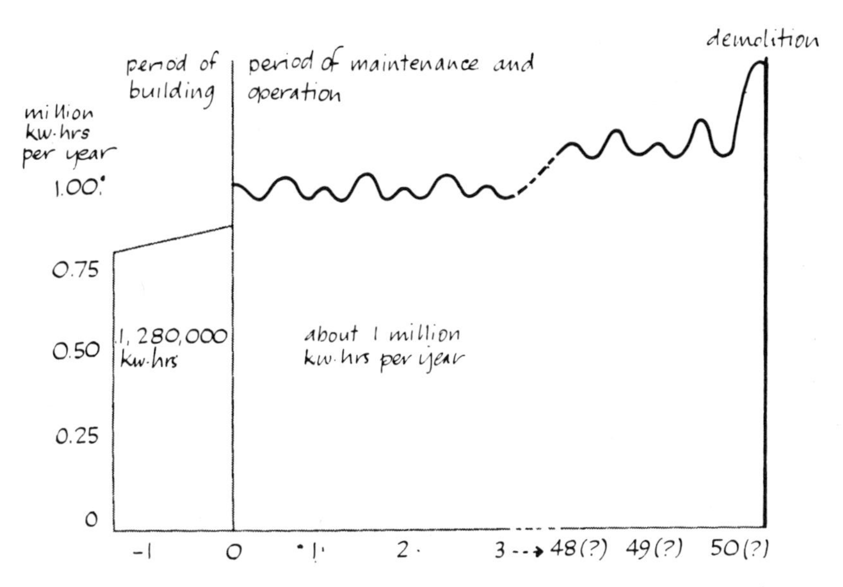

Figure 3 Characteristic energy utilization curve (per million dollars of building in million kw-hrs per year)

periods. (This is an approximation since the energy required by products and their components can be expended in a time period unrelated to the period of building.) After the building is in use, and during the life of the building, the energy use can be described by an undulating curve which may have a high peak in midsummer and a less high peak in midwinter with troughs at spring and fall. The line is made up of a series of daily peaks and troughs, as well, with peaks during the afternoon in winter and at midday in the summer. The whole undulating line climbs slowly year by year as equipment and assemblies become less efficient and energy is used in replacements. It has a momentary peak at demolition and then drops to zero.

As a broad generalization, for every million dollars of new building, there is now an electrical energy expenditure of 1.28 million kwh. See Table 1. To operate and maintain that million dollars worth of building on an all electric basis requires 1 million kwh per year, rising to 1.13 million kwh per year as maintenance requirements increase. This figure is based on the Energy Study by Flack and Kurtz for a theoretical 1 million square foot building with an assumed cost of $35 million.

We have shown that while the characteristic shape of the curve will not change, we can push down the energy required to build and maintain the building by at least 25% and that this would have a saving in energy at the source of 35%. This figure, carried back into the whole national energy picture, can account for 8% of all energy used for all purposes.

The sum of all curves for all buildings, with their individual variations, describes the share of the energy pool, electrical and other, consumed or committed by the products of the architect's decisions. It represents well over half of all electrical usage. We architects can either reinforce the rapid acceleration of energy use or dramatically reduce its rate of consumption, and, in fact, can help reclaim a significant part of our present capacity.

REFERENCES

The Aluminum Association, 1970. *Architectural aluminum report—Aluminum in curtain wall systems.* April, p. 2.

American Society of Heating, Refrigeration and Air Conditioning Engineers, 1967. *Guide and data book equipment.* New York.

Blackwell, H. R., 1959. Specification of interior illumination levels. *Illum. Eng.*, June.

Butler, P. C., and J. T. Rusumore, 1969. *Perceptual and motor skills.*

Construction Review, Sept. 1971. p. 54. U.S. Dept. of Commerce, Washington, D.C.

Cook, Earl, 1971. The flow of energy in an industrial society. *Sci. Am.*, Sept.

Culberson, O. L., 1971. The consumption of electricity in the United States, June.

Design Newsletter, Oct. 1971. Building design and construction, p. 45.

Edison Electric Institute (EEI), 1970. *Statistical yearbook.*

Electrical World, 1970. 21st Annual Electric Industry Forecast, Sept. 15, 1970. Quoted in Environmental Protection Adm., N.Y., 1971, p. 179.

Environmental Protection Administration, City of New York, 1971. *Toward a rational power policy*, Apr. 1971.

Fischer, Robert E., 1971. New methods for evaluating lighting systems. *Architectural Record*, Oct.

Gordian Associates, Inc., *The Comparative environmental impact in 1980 of fossil fuel space heating systems versus electric space heating.* Electric Energy Association, New York, Mar. 1972.

GTE Sylvania, 1971. Advertisement in *Building Design and Construction.* pp. 16–17, Oct.

Hebb, D. O., 1972. *Textbook of psychology*. Philadelphia: W. B. Saunders Co., p. 210.

National Board of Fire Underwriters, 1955. National building code. New York City Planning Commission, *Plan for New York City, critical issues.*

St. Joe Minerals Corporation, 1971. Advertisement in *Eng. News Record,* Nov. 4, p. 38.

Sampson, F. K., 1970. *Contrast rendition in school lighting.* Education Facilities Laboratories.

Scientific American, 1970. The input/output structure of the United States economy.

Steiner, Sheldon, Illumination consultant for the installation of new lighting in St. Patrick's Cathedral, N.Y.

Thompson, F., 1970. *Elec. Comfort Cond. J.,* Apr., pp. 13–16. Quoted in Culberson, 1971, p. 19.

Tinker, M. A., 1939. The effect of illumination intensities upon fatigue in reading. *J. of Educ. Psych.*

U.S. Statistical Abstract, 1970.

Westinghouse Electric Corporation, 1960. *Lighting handbook.* Data on luminaire efficiency extrapolated from this book.

MICHAEL PERELMAN

Mechanization, Energy and Agriculture

American agricultural technology is extremely dependent upon an abundance of cheap energy. So long as a continued supply of cheap energy is forthcoming, it may be able to proceed in the same direction as in the past; but, should it have to face an energy crisis in the future, or even a substantial rise in the price of energy, agriculture may have to undergo a drastic reorganization. This chapter investigates the nature and degree of agriculture's dependency on cheap energy with the hope that it will lead others to think about what an energy crisis would mean for agriculture. United States agricultural policy makers should not be caught unprepared should such a contingency arise.

The harnessing of fossil fuels is the major reason for the spurt in productivity per man in agriculture in the 20th century.

As late as 1920, more than 20 million horse-power was provided by horses and mules which had to be fed from the land (Fox, 1966). If U.S. farmers today were to get all their farm horse-power from horses, they would need to use twenty to fifty times as much cropland for grazing as was used in 1920, should they continue to farm with as much horse-power as is now used (Barnett, 1967). With the adoption of the tractor, this land was freed to produce food for humans instead of horses and mules since the tractor feeds on oil. Not only is land freed by the tractor; labor is also freed because one man plowing with a tractor can do the work of several men plowing with mules. The replacement of human energy by mechanical power is shown in Table 1.

Mechanical power has several advantages over animal power. First, mechanical power permits one man to farm a larger area than he could manage with animal power, and, given the low price of fuel, his expenses increase only marginally when he extends his acreage.

Table 1 Non-Animate Horse-Power and Man Hours Worked in U.S. Agriculture 1920–1969

Year	Tractor Horse-power Millions	Cost of Operating and Maintaining Farm Capital Million Dollars	Man Hours of Farm Work Millions
1920	5	—	13,406
1950	93	5,640	6,922
1960	154	8,310	4,590
1969	204	11,500	3,431

Source: Changes in Farm Production and Efficiency, A Summary Report, 1970, United States Department of Agriculture, Statistical Bulletin, No. 233, Washington, June, 1970.

The second advantage of mechanization is related to the division of labor in agriculture. Earlier economists were not very optimistic about improvements in agricultural productivity since they believed that such improvements depended upon the division of labor and that there was not much scope for improvement in the division of labor on the farm. This point was made twice by Adam Smith in his Wealth of Nations and was echoed by John Stuart Mill who wrote (Mill, 1909):

> Agriculture . . . is not susceptible of so great a division of occupations as many branches of manufacturers, because its different operations cannot possibly be simultaneous. One man cannot always be ploughing, another sowing, and another reaping. A workman who only practiced one agricultural operation would lie idle eleven months of the year. The same person may perform them all in succession, and have, in most climates, a considerable amount of unoccupied time.

Alfred Marshall also accepted the difficulty of the industrialization of agriculture which he said "cannot move fast in the direction of the methods of manufacturing" (Owen, 1966).

But, in fact, agriculture did move quickly in the direction of manufacturing since the mechanization of agriculture meant that many jobs were transferred from the farm to the factory. For instance, the growing of hay for horses was replaced by the refining of petroleum for tractors. In other words, mechanization of agriculture created a necessity for more non-farm inputs in agriculture. Workers who were displaced by the new machines migrated to the city where many of them were employed in producing machines and other non-farm inputs for their comrades who remained on the farm. Between 1919 and the present, U.S. industries and the service sector have employed almost two million man-years per annum in the production of goods and services used in American agriculture (Dovring, 1967).

In their new employment, these men worked on monotonous assembly lines where the division of labor was more advanced than on the farm. The seasons of nature no longer dictated the pattern of their work; rather, they conformed to the more regular rhythms of the machine. So, in effect, much of the farm work was transported from the farm to the city where it was performed like any other industrial process. The efficiency of the industrial process lowered the costs of these products to the point where the capitalists were able to earn a substantial profit while selling them to the farmer cheaply enough to make them economical.

Until the harnessing of fossil fuels on the farm, most of the increases in the productivity of agricultural labor could be explained by increases in capital and land. With the advent of the tractor on the farm, productivity rose faster than capital (USDA, 1970). In fact, productivity per man rose much faster than in manufacturing (Chandler, 1962; Lave, 1962).[1] Much of this increase in productivity can be attributed to increases in inputs purchased by farmers (Meiburg, 1962).

ENERGY USE IN AGRICULTURE

To show what high levels of energy consumption mean for agriculture, Cottrell (1955) tried to compare the energy budgets of Japanese and American farming. He found comparable statistics from two rice farms —one in Japan and the other in Arkansas. Each had approximately the same yield per acre, although the yield per man in the U.S. was higher. In Japan, an acre could be cultivated and harvested with about 90 man-days, which is equivalent to 90 horse-power hours. On the Arkansas farm, more than 1,000 horse-power hours of energy per acre were used just to power the tractor and truck.

Moreover, the non-residential consumption of electrical energy exceeded 600 horse-power hours per acre. Cottrell did not calculate energy required to produce the tractors and equipment.

On a national level, U.S. farmers use about eight billion gallons of fuel each year to run their tractors.[2] This consumption represents

[1]Farm productivity per man may now be leveling off. The USDA Index of Agricultural Productivity stood at 83 back in 1955. During the next five years, it rose to 96 and, by 1965, it had reached 101. Between 1965 and 1970, it had fallen to 99 (USDA, 1971).

[2]In the U.S., there are about 5 million tractors with an average size of 40 horse-power (USDA, 1970). Although tractors were used an average of 605 hours annually in 1956 (USDA, 1960), unpublished data show that the annual usage has fallen to about 400 hours per year (Strickler, pers. comm.). A tractor consumes about 0.1 gallon of fuel for each horse-power hour of use (Shultis, 1961). Thus, each tractor represents 16,000 horse-power hours of use annually which indicates about 1600 gallons of fuel. Five million tractors each using 1600 gallons of fuel

about 40 gallons of gasoline for every American. Since a gallon of gasoline contains about 126,000 BTUs, farm tractors use 5,040,000 BTUs per person in the United States. The average American consumes about 3000 calories of food daily which adds up to an annual consumption of 4,380,000 BTUs; that is, tractors burn up more energy than is contained in the food we eat.

The energy consumed by tractors is only the beginning of the energy input required by present-day American farming techniques. The fertilizer industry consumes enormous amounts of energy. Current technology requires about 2.67×10^7 BTUs per ton of nitrogen fertilizer produced commercially (Delwiche, 1968). In 1969, U.S. farms consumed about 7.5 million tons of nitrogen fertilizer which required about 2×10^{14} BTUs, or about one million BTUs for each American. But then, nitrogen fertilizer makes up only one-fifth of our total commercial fertilizer supply (USDA, 1971). A. B. Makhijani estimates that the over-all average energy use in the fertilizer industry is a little less than 2×10^7 BTUs per ton of fertilizer (Makhijani, pers. comm.). Since the total 1969 fertilizer usage was about 40 million tons, Makhijani's figures represent a total of about 8×10^{14} BTUs, or more than four million BTUs for every American.

The energy costs of farm chemicals are substantial. The Economic Research Service estimates indicate that about four billion gallons of petroleum are used for agricultural petrochemicals each year (USDA, 1968).

Electricity also contributes a great deal to farm production. In 1970, U.S. farms consumed more than 50 trillion BTUs of electrical energy (Hirst, pers. comm.). Since the production of one BTU of electrical energy requires about 3.07 BTUs of fuel input, electric power production for farm uses actually represents more than 150 trillion BTUs per year, or an equivalent of about 750 thousand BTUs of resource energy yearly for every American.

The production of farm equipment also consumes much energy. Each year, the farm implement industry alone uses about 520 thousand BTUs for every American, not counting the energy used by the supporting industries which supply that industry (Makhijani and Lichtenberg, 1971).

Much of the energy used in the distribution and processing of food should also be charged to the organization of agricultural production which has minimized production costs through regional specialization.

---

represent 8 billion gallons of fuel. One study by the Economic Research Service indicates that tractors only consumed 3.5 billion gallons of petroleum fuel in 1965 (USDA, 1968). This same study estimates that total farm consumption of fuel was about 9 billion gallons in 1965.

This specialization requires that food be transported longer distances and also that much food be processed to avoid spoilage in the often circuitous road from farmer to consumer. In 1969, almost $5 billion was spent simply for transporting food by rail and inter-city trucks (USDA, 1971).

The energy cost of the food processing sector is also significant. Makhijani estimates that this sector consumes about 10^{15} BTUs per year, or an amount comparable to the consumption of energy by tractors (Makhijani, pers. comm.).

EFFICIENCY

If efficiency is measured in terms of the conservation of energy, then American agriculture comes out very poorly. Harris estimated that traditional Chinese wet rice agriculture at its best could produce 53.3 BTUs of energy for each BTU of human energy expended in farming it (Rappaport, 1967). But this energy came from people who burnt rice in their bodies, rather than fossil fuel in tractors.

Delwiche estimates that 1.5×10^9 calories were used for cultivation of each hectare (Delwiche, 1968) in the U.S.; this translates into 2.4 million BTUs per acre with about 300 million acres of cropland farmed each year (USDA, 1971), not including grazing land. Food production would require about 720×10^{12} BTUs, or about 3.6 million BTUs for each American we feed. These figures indicate that U.S. food agriculture consumes about as much energy as it produces and Delwiche's estimate does not even take into account the energy required to produce the farm equipment or to store and distribute the food! Delwiche suggests conservatively doubling the energy input to account for these units. Thereby, food production would appear to consume twice as much energy as it yields in actuality.

Another method of estimating the energy cost of agriculture would be to add together the energy cost of operating tractors, the energy cost of producing the pesticides, electricity and farm implements and the energy cost of the food processing industry. These activities require about 14 million BTUs per year for every American, or three times the amount of energy consumed as calories at the table, in spite of the energy costs which are excluded from this calculation. The magnitude of the excluded costs can be envisioned by these examples: farmers purchase products containing 360 million pounds of rubber, about 7% of the total U.S. rubber production, and 6.5 million tons of steel in the form of trucks, farm machinery and fences. Farms consume about one-third as much steel as the automotive industry (Committee on Agriculture, 1971).

In contrast, for each unit of energy the wet rice farmer expends, he can get more than 50 in return; for each unit of fossil fuel energy we expend in food production, we get about one unit in return. On the basis of these two ratios, we can say that, in terms of energy expended over and above that required to sustain the life of the farmer, Chinese wet rice agriculture, at its best, is about 50 times as efficient as average food production in the U.S., and that primitive agriculture in general is ten to thirty times as efficient. This measure of efficiency, however, deals only with energy inputs and outputs and does not consider such important variables as output per man or per acre.

CONCLUSION

Of course, agriculture is not the main user of energy in our society. In 1970, the U.S. consumed about 64,000 trillion BTUs of energy (Cook, 1971), or 320 million BTUs per capita per year. For instance, a typical American consumes about 1.3 million BTUs annually just to watch a black and white television set (Brune, 1972). By that standard, agriculture's consumption of 14 million BTUs per year to feed one person does not seem extravagant. Besides, the U.S. uses more than 30% of its acreage for exports which feed citizens of other nations and some of our crops are used for non-food purposes (USDA, 1971). The problem is that agriculture traditionally has been, and it has been proposed that it could again become, an *energy producing* sector of the economy.[3] Harvested crops capture solar energy and store it as food or some other useful product. Yet, the energy captured is small compared to the energy we burn to capture it. Agriculture, as a result, has become a major consumer of our stores of energy, using more petroleum than any other single industry (Committee on Agriculture, 1971).

If the world is facing a future with rising energy prices, the highly mechanized technology currently used in U.S. agriculture may be in-

[3]Some writers argue that crops have the potential of becoming a major energy source for the economy. Kramer et al. point out that corn silage grown as feed costs about $7.50 per ton undelivered in 1971, and that the energy stored in this corn is equivalent to that of coal at $12.00 per ton (Kramer et al., 1972). Of course, the cost of corn silage "fuel" could be expected to rise if production were increased many-fold in order to reach a point where the corn harvest could supply a sizable fraction of our energy needs. Besides, fossil fuels are a much more concentrated form of energy which makes their transport simpler. However, as one writer has pointed out, much of the "decline of (harvested) materials before the onslaught of synthetic products in recent years is due to the enormous amount of scientific and technical research that has been carried out to ensure the fullest utilization of petroleum by-products. No comparable effort has been made for plant products . . ." (Forthomme, 1968).

appropriate. Moreover, if such a future is coupled with large increases in population, this form of technology may prove inadequate to provide the yields necessary to feed the world's population. Wheat yields per acre of U.S. farms are only slightly higher than those of Nepal and half as large as those of Denmark. Western European rye yields are about half again as large as those of the U.S. and both Japan and Spain grow more rice per acre than the U.S. United States oats yields are more than doubled in the United Kingdom, and New Zealand and the Common Market's yields on an American specialty, corn, are about 10% higher than that of the U.S. (USDA, 1971; Lave, 1962; Tweeten and Tyner, 1965). In fact, American farms produce less value of crops per acre than those of Southern or Eastern Asia or Central America (Tindall, 1972).

REFERENCES

Barnett, Harold J., 1967. The myth of our vanishing resources. *Trans-Action* June: 7–10.

Brune, W. D., Jr., 1972. The economic impact of electric power development. A talk before the National Engineers Week Symposium, Chico State College, Feb. 26.

Chandler, Cleveland A., 1962. The relative contribution of capital intensity and productivity to changes in output and income in the United States economy, farm and non-farm sectors. *J. Farm Economics.* 44 (May): 335–348.

Cottrell, Fred, 1955. *Energy and society.* McGraw-Hill, New York: McGraw-Hill, esp. pp. 138–140.

Committee on Agriculture, House of Representatives, 1971. *Food costs-farm prices: A compilation of information relating to agriculture.* U. S. Government Printing Office, 92nd Congress, 1st Session, Washington, D.C., July 1.

Cook, Earl, 1971. The flow of energy in an industrial society. The editors of *Scientific American, Energy and Power,* San Francisco: W. H. Freeman.

Delwiche, C. C., 1968. Nitrogen and future food requirements, Research for the world food crisis. A Symposium Presented at the Dallas Meeting of the American Association for the Advancement of Science, December, Daniel G. Aldrich, Jr., ed., Publication 92, Washington, D.C.

Dovring, Folke, 1967. *The productivity of labor in agricultural production.* Univ. of Illinois College of Agriculture, Agricultural Experiment Station Bulletin 726, Sept.

Forthomme, P. A., 1968. Can rice replace petroleum. *Ceres* 1, no. 5 (Sept.-Oct.): 50–51.

Fox, Austin, 1966, *Demand for farm tractors in the United States.* U. S. Department of Agriculture, Economics Research Service, Agricultural Economic Report No. 103, Nov.

Hirst, Eric. National Science Foundation, Environmental Program, Oak Ridge National Laboratory.

Kramer, Marc, et al., 1972. Solar energy. In the American Association for the Advancement of Science, Committee on Environmental Alterations, *Electric power consumption and human welfare.*

Lave, L. B., 1962. Empirical estimates of technological change in the United States agriculture. *J. of Farm Economics* 44, (Nov.): 941–952.

Makhijani, A. B., and A. J. Lichtenberg, 1971. An assessment of energy and materials utilization in the U.S.A. Memorandum No. ERL-M310, Electronics Research Laboratory, College of Engineering, Univ. of California at Berkeley, Sept. 22.

Meiburg, Charles O., 1962. Nonfarm Inputs as a Source of Agricultural Productivity, *Food Res. Inst. Studies* 3, No. 3 (Nov.): 297–321.

Mill, John Stuart, 1909. *Principles of political economy,* ed. W. J. Ashley. London, pp. 131–132.

Owen, Wyn F., 1966, The double devolopmental squeeze on agriculture. *Am. Economic Rev.* 56, no. 1 (Mar.): 43–70.

Rappaport, Roy A., 1967. *Pigs for the ancestors.* New Haven, Conn. Yale University Press, p. 262.

Shultis, Arthur. Estimating tractor costs, 1961. In *Data for your farm management handbook,* No. 2, July, reissued September, 1963, Univ. of California, Agricultural Extension Service.

Strickler, Paul. Agricultural economist with the Farm Production Economics Division of the United States Department of Agriculture, personal communication.

Tindall, Diane, 1972. Five years later. In *War on Hunger* 6, 10 (Oct.): 1-5. U. S. Department of State, Agency for International Development.

Tweeten, Luther G., and Fred H. Tyner, 1965. Toward an optimal rate of technological change. *J. Farm Economics* 46, no. 5 (Dec.): 1075–1084.

United States Department of Agriculture, 1971. *Agricultural statistics, 1971.* U. S. Government Printing Office, Washington, D.C.

United States Department of Agriculture, Agricultural Research Service, 1960. *Farm tractors: Trends in type, size, age and use.* Agricultural Information Bulletin, Aug.

United States Department of Agriculture, Economic Research Service, 1968. *Structure of six farm input industries.* Washington, D.C.

———, 1970. *Changes in farm production and efficiency, 1970.* Statistical Bulletin, No. 233, Washington, D.C., June.

BARRY COMMONER AND MICHAEL CORR

Power Consumption and Human Welfare in Industry, Commerce, and the Home

The extensive social costs inevitably involved in the present rate of growth of power production indicate a strong need for consideration of the problem of reducing the demand for power. Since electricity in itself is of no direct use to man, but becomes valuable only when converted to some good, whenever the efficiency with which electricity is converted into a good is less than optimal there is an opportunity to reduce power consumption at no sacrifice of human welfare. However, it is apparent from what follows here that most efforts to improve the efficiency with which power is used have very serious economic and social consequences.

To begin with it is important to consider the relationship between power consumption and the standard of living—or the goods available per capita. Since it has been observed that for various countries, GNP per capita is roughly proportional to energy consumed per capita (Cambel et al., 1965), it is sometimes assumed that goods production is proportional to GNP and that electric power consumption is proportional to total energy consumption; hence, that

$$P = k \cdot G$$

where P is the amount of power consumed annually, G is the amount of goods produced annually, and k is a factor which has the dimensions of goods produced/power consumed. The proportionality factor, k, would then measure the efficiency with which power is used to produce the goods. That the standard of living, or goods consumed per capita, is proportional to power consumption is thus often taken to imply that reduction in power consumption can be accomplished only by reducing the standard of living, the size of the population, or both. This assumes, however, that the proportionality factor, goods produced/power consumed, is constant; clearly if the value of this factor is increased it becomes possible to reduce power consumption at no sacrifice in goods produced, and without reducing the size of the population.

Another preliminary observation is relevant here: reduction of power production is likely to be a far more effective way of reducing the resultant environmental degradation than any conceivable method of controlling the emission of pollutants.

Thus, it becomes important to look into the elasticity of the factor k, that is, to determine the degree to which the efficiency with which power is used can be improved.

The basic course of the utilization of electric power in the U.S. in the postwar period is shown in Figure 1, which is taken from an excellent report by the Illinois Geological Survey (Risser, 1970). The familiar exponential growth in power consumption—which in the last five years has increased at the rate of 4.9% *per capita* annually—is evident.

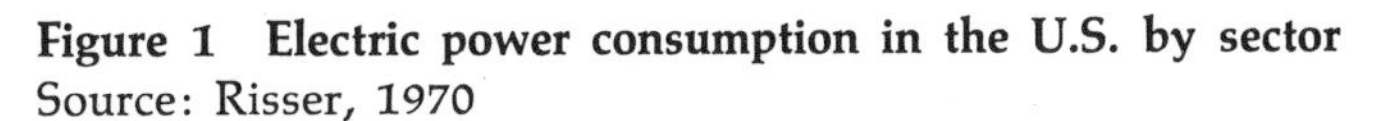

Figure 1 Electric power consumption in the U.S. by sector
Source: Risser, 1970

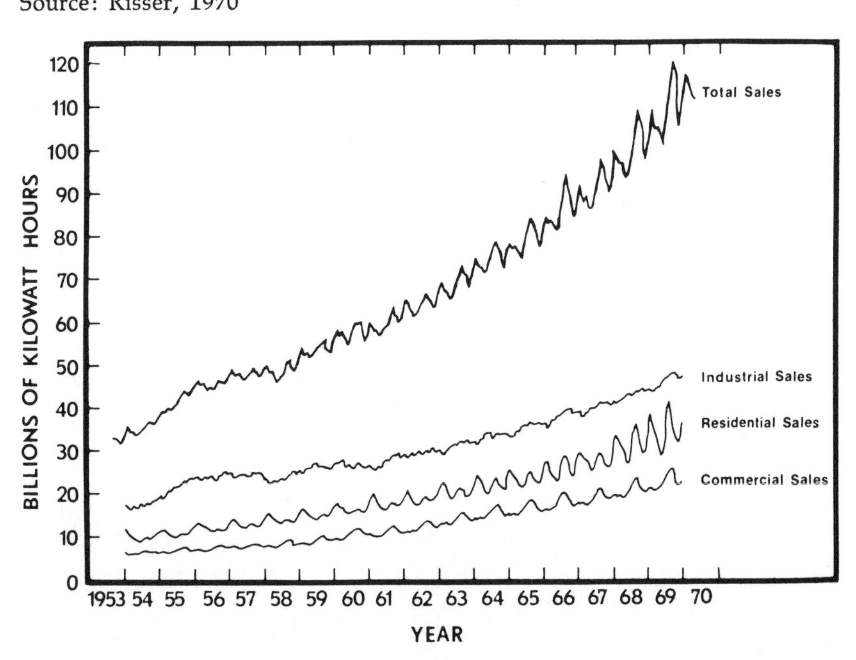

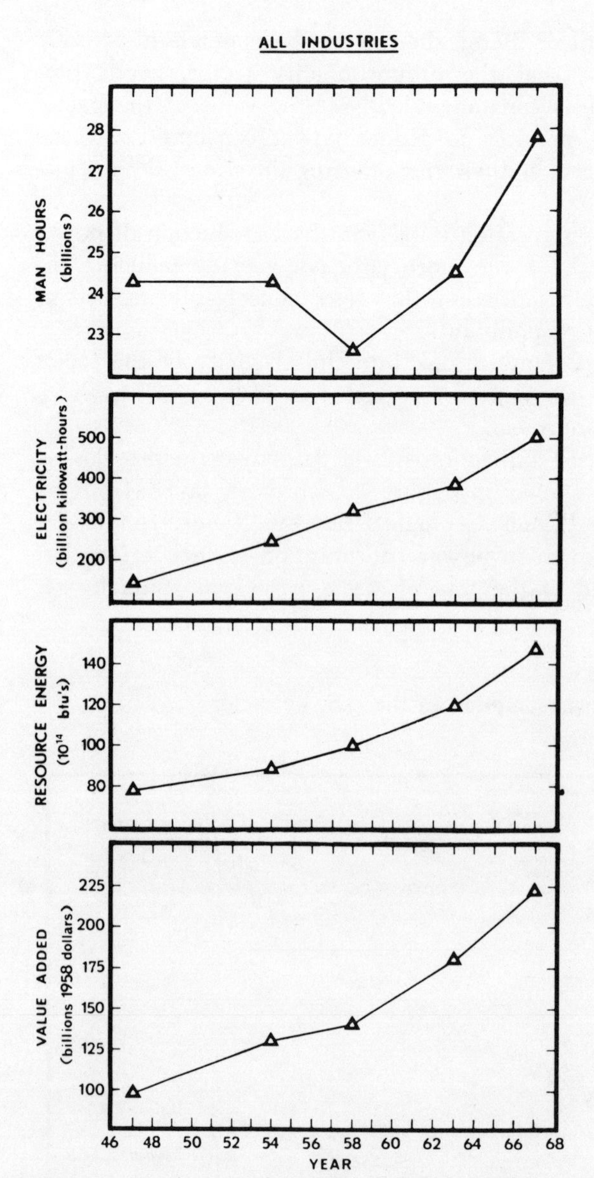

Figure 2 Use of electrical power by industries
Source: Data from U.S. Dept. Comm., 1971. Man hours and value added, Vol. 1, p. 26. Electricity and Resource Energy (except Resource Energy, 1947) SR4, pp. 8, 9. Resource Energy 1947, from: Lyon, 1951.

INDUSTRIAL POWER CONSUMPTION

The use of electric power by industry has increased exponentially, with a doubling rate of about 14 years at present. This is shown in Figure 2, along with other relevant parameters regarding industrial production. Industry is, of course, a major means of converting power into useful goods and, given the considerations outlined before, it is necessary to relate its consumption of power to the concurrent production of goods. This is, naturally measured in economic terms. A convenient and readily available measure of the economic good resulting from an industrial operation is *value added*. This is defined as the value of the goods shipped by industry minus the sum of the costs of materials, fuel and power used, and contract services. If, in turn, the cost of labor and new capital are subtracted from value added, one arrives at the gross return to the entrepreneur, representing profit, before taxes and before the subtraction of other non-productive expenses, such as advertising.

Figure 2 shows that since 1947, for total U.S. industrial production, value added and electricity consumed have risen exponentially but at somewhat different annual rates. Overall, value added has increased about 2.3-fold, from about $98 billion to about $222 billion (all these figures are computed to 1958 dollars to compensate for inflation); electricity consumed has increased about 3.6-fold, from about 141 billion kwh to about 506 billion kwh. Total resource energy used in industry—that is, the energy content of all fuel used in industry, including that needed to produce electricity—has about doubled in that time. Labor employed in industry follows a distinctly different course; total man-hours expended annually in U.S. industrial activity increased only 15%, from 24.3 billion in 1947 to 27.8 billion in 1967.

There is a close and useful analogy between the roles of labor and electric power in industrial production. Both are non-storable entities, which become valuable only in their use; both are consumed in the course of the process in which they are engaged. There is also a close functional relationship between the roles of labor and electric power in the productive enterprise, since electric power is the most convenient means of substituting for, or amplifying, the muscular power and manipulative capabilities of human beings.

The economic value of labor is usually given by the term *labor productivity*[1] which may be defined as value added per man-hour of

[1]In general, if one wishes to evaluate the contribution of a production factor, such as labor or power to the overall value of a product, it is necessary to compute not only the labor or power expended in the final production process, but also that expended in the entire sequence of productive activities which leads up to the final one. This includes: labor or power expended in mining ores, in

labor. By analogy with labor productivity, then, we may define the *power productivity* of an industrial enterprise as the quotient value added/electricity consumed. Figure 3 B and C shows that both labor productivity and power productivity for all U.S. industry have exhibited striking changes since 1947, but in opposite directions. There has been a continued increase in labor productivity (although the rate of increase has been declining in recent years): the overall change is about 2-fold in the 20-year period. In contrast, power productivity declined sharply between 1947 and 1958, remaining more or less constant since then. Overall there has been a 35% decline in industrial power productivity since 1947.

When a comparable analysis is made of the changes since 1947 in the productivity of *resource energy consumed,*[2] (i.e., dollars of value added per million BTUs of resource energy, where resource energy is the total amount of fuel consumed to produce heat and power for industry), the results shown in Figure 3 are obtained. Value added per unit resource energy increased between 1947 and 1954, and has since then remained fairly constant at about $15 per million BTUs. The 30% increase in resource energy productivity appears to be due to rapid relative growth in the mid-productivity industries (Table 1, Group B). However, since 1954, in most industries resource energy productivity has been declining, although the decline is not as rapid as that of power productivity.

Electric power consumption has increased at a rate considerably in excess of the rate of increase in the economic benefits yielded by industrial production. A useful way to visualize this effect is to compute how electric power demand would have increased after 1947 if power productivity had remained constant rather than declining. Such a computed curve (Fig. 4) shows that if the technological transformations in the use of electric power which took place in U.S. industry after 1947 had *not* occurred, industrial power consumption in 1969 would have been reduced by about 35%.

---

converting them to metals which serve as materials in the final productive process, in the transportation needed at various stages, in the necessary administrative operations, etc.; these data are available from economic input-output tables.

[2]Resource energy, an estimate of the BTU equivalent of all fuels used for energy—including those used in power generation—is computed using these conversion factors:

1 kwh electricity requires 3.07 kwh fuel input
1 kwh = 3412 BTU.

Therefore 1 kwh requires 3.07 × 3412 BTUs resource energy = 10,500 BTU. Resource Energy (BTU) = 3412 (Total purchased fuels and electric energy in kwh from Census of Manufacturers + 2.07 electric energy purchased in kwh from Census of Manufactures).

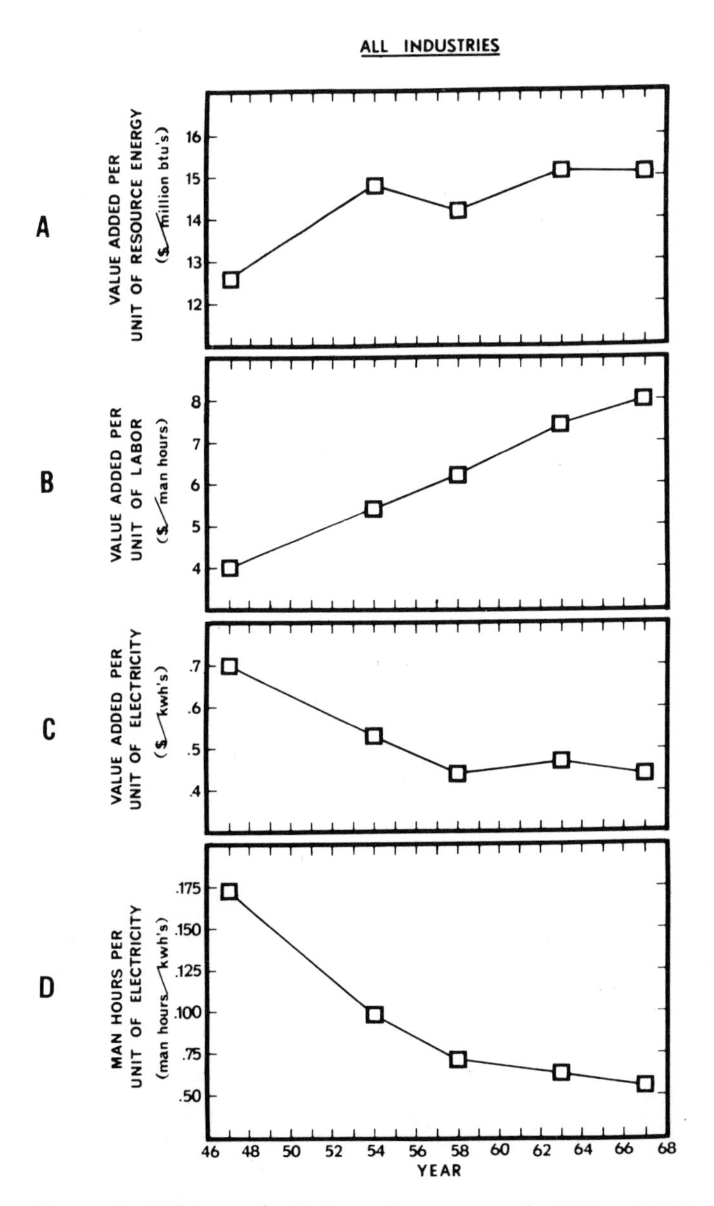

Figure 3 Labor productivity and power productivity of U.S. industry
Source: Computed from data in Figure 2.

Table 1 Power Productivity and Consumption by Each Industrial Sector for 1947 and 1967.

SIC 2-digit categories[a]	Value Added 1958 Dollars x 10^9		Electric Power Used Kwh x 10^9		Electric Power Productivity 1958 Dollars/Kwh		Man-Hour Productivity 1958 Dollars/MH		Resource Energy Productivity 1958 Dollars/10^9 BTU		
	1947	1967	1947	1967	1947	1967	1947	1967	1947	1954	1967
23 Apparel	5.87	8.53	0.85	3.61	6.91	2.36	3.24	3.92		173	119
21 Tobacco Mfg.	0.85	1.73	0.16	0.85	5.26	2.03	4.28	13.72		79	71
27 Printing & Publication	5.62	12.17	1.28	5.82	4.39	2.09	6.33	10.18		173	115
31 Leather & Prod.	2.03	2.23	0.57	1.33	3.54	1.67	3.00	4.06		55	53
38 Instruments	1.51	5.44	0.55	3.08	2.77	1.77	3.87	10.27		74	78
25 Furniture & Fixtures	1.78	3.54	0.83	2.52	2.16	1.40	3.05	4.94		55	56
Group A	17.66	33.64	4.24	17.21	4.17	1.95	3.88	6.35		105	89
35 Machinery	10.36	23.61	5.92	17.26	1.76	1.37	4.00	8.48		57	56
34 Fabr. Metal Prod.	6.51	15.30	3.90	14.76	1.67	1.04	3.84	7.08	38	39	38
24 Lumber Wood Prod.	3.33	4.22	2.34	7.97	1.43	0.53	2.66	4.32		29	18
36 Elec. Equip.	5.11	20.77	3.62	19.20	1.41	1.08	4.00	7.95		46	57
37 Trans. Equip.	7.73	23.89	6.06	23.56	1.28	1.01	3.94	8.70		48	45
20 Food & Kd.	12.06	22.57	10.18	26.79	1.18	0.84	5.09	9.99	18	17	21
Group B	45.10	110.36	32.02	109.54	1.41	1.01	4.05	8.15		33	37

22 Textiles	7.04	6.91	10.04	20.80	0.70	0.33	3.05	4.09	21	16	15
30 Rubber & Plastic Prod.	1.72	5.77	3.45	10.77	0.50	0.54	4.05	7.07		15	23
32 Stone, Clay, Glass Prod.	3.04	7.07	8.02	20.81	0.48	0.36	3.63	7.45	3	4	5
29 Petroleum & Coal Prod.	2.63	4.60	6.50	22.28	0.41	0.21	7.44	22.78		4	3
28 Chemicals	7.03	19.97	19.61	116.83	0.36	0.17	7.21	18.39	7	7	6
26 Paper & Allied Prod.	3.85	8.27	15.39	49.07	0.25	0.17	4.50	7.72	6	7	6
33 Primary Metals	7.58	16.94	40.65	131.95	0.19	0.13	3.69	8.11	2	5	5
Group C	32.89	69.53	103.66	372.51	0.32	0.19	4.21	8.80		6	6
Manufacturing	95.65	213.53	139.92	499.26	0.68	0.43	4.07	7.99	11.2	14.4	14.6

[a] Dept. of Commerce, Census of Manufacturers Standard Industrial Classification System (SIC)—2 digit categories.

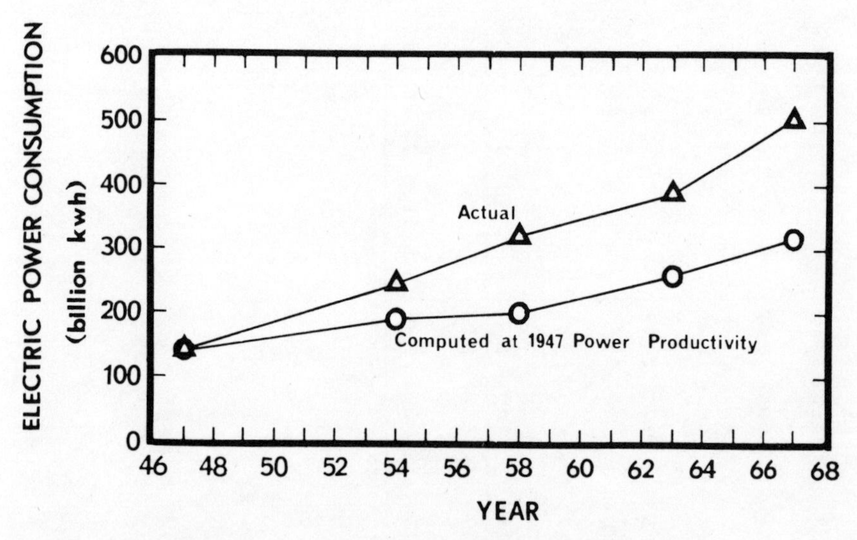

Figure 4 Industrial power consumption (data from figure 2)
computed power consumption for year X =

$$\frac{\text{1947 power consumption} \times \text{value added for year X}}{\text{1947 value added}}$$

Since, as shown previously, the power productivity of U.S. industry has not remained constant since 1947, but has in fact declined, it becomes important to determine whether this trend can be reversed, and what the likely consequences of such an undertaking are for society. This requires a more detailed analysis of the *reasons* for this trend toward lower power productivity. For this purpose it is important to examine the trends among the different sectors of industry. The simplest classification of industries provided by the Census of Manufacturers (U.S. Dept. Comm, 1949, 1957, 1961, 1966, 1971) is that represented by 2-digit Standard Industrial Classification (SIC) numbers, as enumerated in Table 1. For each of the sectors, at approximately 5-year intervals, the Census of Manufacturers reports data from which one can compute the power productivity and total consumption of electric power of each industrial sector.

Figure 5 shows how different industrial sectors have changed, in respect to power productivity and power consumption, between 1947 and 1967. First it is evident that nearly all sectors have declined in power productivity in that period. In 1947 the highest power productivity (about $6.90 of value added per kwh) was that of the apparel industry. In 1967 that industry was still highest in power productivity, but at the level of $2.36 of value added per kwh. At the same time, partly because of the decline in power productivity and partly because

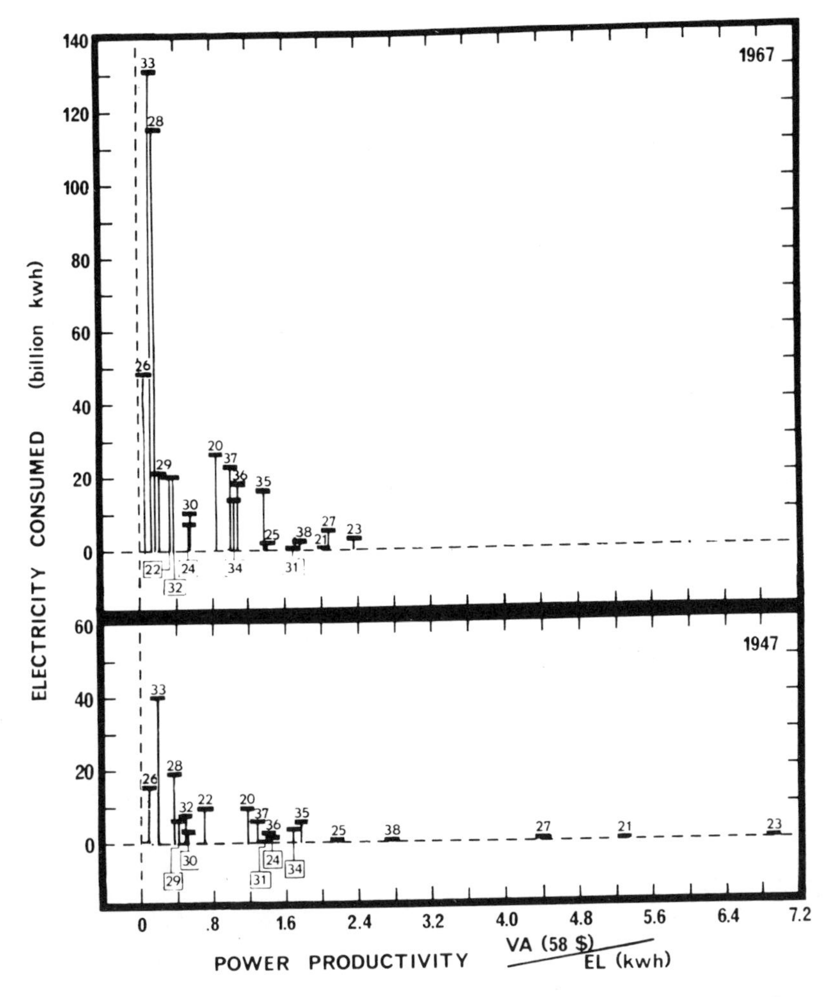

Figure 5 Power productivity and electric power consumption by 2-digit SIC classification 1947 and 1967

Source: Data from Dept. Comm. 1971, 1949

of the increase in total production between 1947 and 1967, the total amount of electricity consumed by that industry increased several-fold. Nearly every industrial sector exhibits these trends.

A second feature illustrated by Figure 5 is that power consumption by the process industries, which have very low power productivities—primary metals (SIC 33), paper and allied products (SIC 26), chemicals (SIC 28), petroleum and coal products (SIC 29), etc.—contributed much more to the absolute growth of power consumption than group A or

B (Table 1). Table 1 shows the rate of growth in power consumption of three groups of industries, classified according to power productivity in 1947. The group with the lowest power productivity ($0.19 to $0.70/kwh) contributed 75% of the growth in power consumption by all manufacturing between 1947 and 1967. However, the contribution of this group of industries to the growth in value added from all manufacturing during 1947 to 1967 was only 27%. The contribution to growth in national power consumption by the groups of industries with power productivities between $0.70 and $1.76 was only 22%. However, this group of industries contributed 48%, by far the largest share to the growth in total value added by industry from the three groups. Finally, the group of industries with power productivities above $1.76 contributed only 4% to the growth in power consumption while it contributed 12% to the growth in value added by manufacturing.

On the basis of this analysis it becomes evident that the rapid growth in industrial power consumption has not paralleled industry by industry growth in the value of the goods produced by each industry, in large part because only a few industries with low power productivities, such as chemicals and primary metals, or those which have experienced immense growth in their contribution to value added, such as transportation and electric equipment, account for a good deal of the growth in power consumption. Analysis of growth in power consumption by manufacturing is thus complicated by differences in power productivity superimposed on great differences in the growth of their contribution to value added by manufacturing.

Similar relations are evident *within* a single industrial group, such as primary metals. As shown in Figure 6, the major contributors to this group, steel and nonferrous metals (chiefly aluminum), differ considerably in their power productivity. For example, in 1967, the power productivity of steel production was $0.183/kwh, while that of aluminum was $0.13/kwh. Figure 6 also shows that the nonferrous metals, which contribute a significantly smaller share than steel to the total value added by primary metals industries, now consume the largest share of electricity in the group. Again, we find that the industrial activities which are least efficient with respect to the use of power contribute disproportionately to the rapid growth of electricity consumed by industry.

These data illustrate another important trend: the tendency of industries which operate at low power productivities to displace industries which operate at high power productivities. Thus, production of nonferrous metals, especially aluminum, has grown much faster than steel production, largely because of the replacement of steel (and lumber) products by aluminum ones. In the same way, the growth of

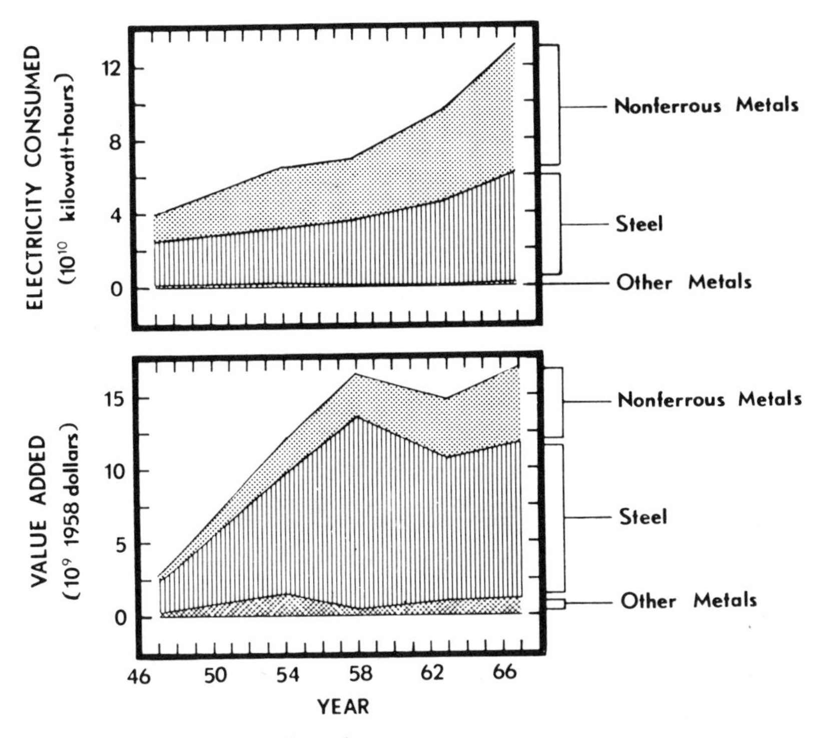

Figure 6 Primary metals industry
Source: Data from Dept. Comm. 1971 SR4, 1966 Vol. 1, 1961 Vol. 1, 1957 Vol. 1, 1949 Vol. 1. Steel figures are SIC 331 & SIC 332. Non ferrous figures are SIC 333, SIC 334, SIC 335, & SIC 336. Other is SIC 339.

the chemical industry—which has a very low power productivity—is largely based on the displacement of a number of natural products, which involve very little power consumption (such as cotton, wood, lumber, and soap made from fat) by synthetic chemical products (synthetic fibers, plastics, detergents).

Thus a good deal of the growth in industrial power consumption for materials production is due to the expansion of the role of power consumptive materials as much as to the expansion of industrial activity. Since we are concerned with the elasticity of this process, especially the possibility of reversing it, it is important to inquire as to whether these displacements were necessary, because, for example, of the depletion of raw materials. Clearly, the foregoing displacements were not *forced;* there is no evidence that aluminum has displaced steel because the latter has been in short supply, or that detergents have displaced soap because we have run out of saponifiable fat (we now export more animal fat than the amount needed to replace U.S. detergent

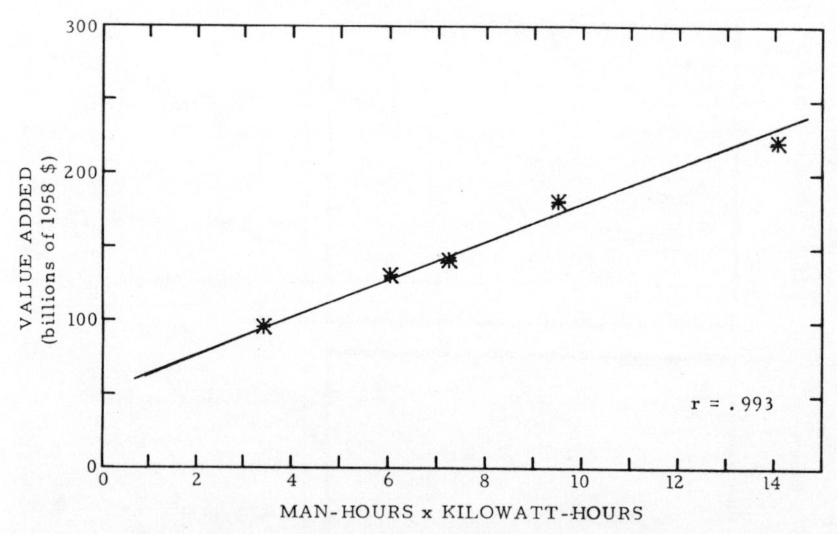

Figure 7 Relationship between value added and the product of kilowatt-hours and man-hours expended in industrial production

consumption with soap). In other words, the industrial displacements which have decreased the efficiency of industrial power consumption are, at least in principle, *reversible*—so that savings in industrial power consumption *could* be achieved by reversing the trends which have been underway in the postwar period.

Apart from such displacements it is also evident that another reason for the declining power productivity of U.S. industry is the progress of automation—in which hand labor is displaced by machines, nearly always driven by electric power. Thus the sharp decline in power productivity in the apparel industry is obviously due to the considerably increased use of machines in that industry in place of hand labor. This is revealed in the overall statistics for U.S. industry by the close relation between the decline in power productivity and the increase in labor productivity. This is illustrated in Figure 7 which shows a linear relationship between value added and the product of kilowatt-hours and man-hours expended in industrial production. This means that the increased productivity of labor is proportional to the increase in the amount of electricity consumed, and the decrease in productivity of electricity is proportional to the decrease in the number of man-hours employed.[3]

[3]This can be seen from the following. According to Figure 7, VA = K (MHxEL) where VA is value added, MH is man-hours and EL is electricity consumed, and signifies an increment in these values. Hence, (VA/MH) = (K)EL and (VA/EL) = (K)MH, where VA/MH is labor productivity and VA/EL is power productivity.

Again, it is useful to consider the elasticity and reversibility of the increasing consumption of power which is associated with the displacement of labor by powered machinery. Clearly this displacement process was not demanded by the reduced availability of labor, and—apart from the considerable economic consequences, which are discussed later—it could be reversed by the simple expedient of increasing hand labor in place of electric powered operations.

Finally, it is useful to examine a specific product with respect to the power required for manufacturing it. Here the necessary data become considerably more complex and uncertain. Keeping these qualifications in mind, it is nevertheless informative to carry out such an exercise if only to develop a general impression of how an effort to save power might be reflected in the nature of the final product.

Since the automobile industry accounts for a large proportion of U.S. raw materials and energy consumption, it forms a clear illustration. The following list (from hearings before the U.S. Senate Public Works Committee E-Y4, p 96/10:J96) shows the percentage of consumption of various raw materials with end uses in the auto industry:

Resource	% of Consumption in Autos
Steel	20
Aluminum	10
Copper	7
Nickel	13
Lead	50
Rubber	60

Table 2 summarizes some tentative computations of the power required to mine and manufacture the metals used in the production of an average passenger automobile. Note first that the total amount of power for metals (expressed as resource energy) required to produce the vehicle increased from 5931 kwh in 1958, to 6353 kwh in 1966, and to 7123 kwh in 1970. The Table also shows that the chief reason for this change is the sharp increase in the use of aluminum, which has replaced steel in certain car parts, especially in the engine, trim, bumpers, and auxiliary hardware. Since aluminum production operates at a very low power productivity, there is a correspondingly large increase—approximately a doubling—in the amount of power represented by the car's aluminum content. We know, of course, that this trend is reversible—that is, the cars that have little or no aluminum content can be built, for in fact they have been in the past. As an exercise in estimating the power saving that might be possible by

Table 2a Total Resource Energy and Resource Energy for Power Consumed in the Mining and Manufacture of Typical Metals for Automobiles

	Resource Energy Per Ton	R.E. for Power Per Ton	%
Cold-rolled steel	15,720	2,605	16.6
Cast iron	7,530	1,965	26.1
Aluminum Forging	78,645	52,000	66.2
Copper	24,185	9,510	39.4
Zinc	32,730	15,700	48.0
Lead	12,000	1,330	11.1

Except for lead, figures were computed from data provided in Stephen Berry (1972). Resource electric power is 3.39 × delivered power. The total energy figure for lead is from Makhijani, 1971. From the Bureau of Census (1971) we found that 11.1% of the resource energy for lead production (not including mining) was for power production in 1967.

sharply reducing the aluminum content of the automobile, Modification I of the 1970 automobile has been carried out by replacing 90% of the aluminum in the vehicle by an equal volume of steel. When the power requirement for producing the materials for the car is computed we find that it is 4866 kwh, as compared with 7123 kwh for the actual 1970 automobile. In Modification II, the metal content of the 1970 automobile was computed by reducing the size of the car to that of the average 1947 automobile. The 1970 automobile thus modified requires only 3986 kwh—44% less than the actual 1970 automobile. (Although simple substitution of steel for aluminum yields a heavier auto using more fuel per mile, the scaled down modification II auto would use less fuel per mile.) One step further in the automobile manufacturing process, the total power required to produce *all* the parts of an automobile and to assemble the finished product (in 1970) is 2076 kwh, which, when added to the 7123 kwh representing the power needed to produce the metals for parts, yields the total power consumption for a passenger automobile. In 1970 this was 9199 kwh. According to *Auto Facts and Figures,* factory sales of passenger cars was 6,546,817 cars in 1970; thus the total power savings achieved by the above modifications would be 14,800 million kwh and 20,500 million kwh respectively. Adoption of Modification I would save 2.1% of total industrial power consumption or 1.0% of total national consumption; Modification II would save 2.9% of industrial power consumption and 1.3% of the national consumption.[4] While these savings are small in themselves, applied in a similar manner to other metallic manufactured products they could add up to an appreciable reduction in power demand.

[4]See footnote c of Table 2b.

Table 2b Savings in Electric Power Consumption through Economies or Transfers in Metals Use in the Manufacture of an Automobile

				Savings	
Metals Consumed	1958	1966	1970	1970 Modification I[a]	1970 Modification II[b]
Steel (tons)	1.200	1.190	1.150	1.293	1.062
Kwh Resource Electricity (R.E.)[c]	3125	3100	3000	3375	2770
Cast iron (tons)	.312	.305	.300	.300	.246
Kwh R.E.	614	600	590	590	484
Aluminum (tons)	.027	.035	.055	.0055	.0045
Kwh R.E.	1438	1863	2925	293.	239.
Zinc (tons)	.045	.052	.043	.043	.0353
Kwh R.E.	428	495	409	409	336
Copper (tons)	.020	.018	.012	.012	.0094
Kwh R.E.	314	283	188	188	148
Lead (tons)	.009	.009	.008	.008	.0066
Kwh R.E.	12	12	11	11	9
Total R.E./car (kwh)	5,931	6,353	7,123	4,866	3,986

[a]Modification I of the auto materials is achieved by replacing 90% of the 0.055 tons of aluminum in the 1970 car with the same *volume* of steel (0.1432 tons). This leaves 0.0055 tons of aluminum in the vehicle. (In 1947 the average vehicle, including automobiles, trucks and buses, included 0.013 tons of aluminum).

[b]The price in 1958 dollars of a 1970 car is less than that of a 1967 car; However, the latter contains more materials. To make a rough projection back to a 1947 size the ratio of the 1967 to the 1947 prices is applied to the materials of the modification I 1970 car.

Year	1958$/car
1947	1460
1967	1780
1970	1640

[c]Resource electric energy computed at 3.39 × electricity to account for efficiency of generation of 32.8% and 90% for transmission.

This exercise is simply an illustration, on the specific level of a particular product, of the general trend in industry as a whole toward substitution of power-demanding materials such as aluminum for less power-consumption materials such as steel. While the exercise is highly tentative in its numerical details it does suggest that savings of the size indicated by the overall industrial data can in fact be achieved by modifying the design of a specific product. It should be noted that in these hypothetical modifications it is assumed that the power required for the actual *production* of each metal and for assembling the automobile is that actually used in 1970. In other words, in this exercise, the techniques of manufacture (including the relative balance between hand labor and powered machines) remain unchanged; only the relative amounts of the different metals employed are changed. If a reversal in the recent trends of the technique of automobile manufacturing were acceptable—in particular a return to the amount of hand labor characteristic of automobile manufacturing in 1947—an additional reduction of about 12,702 million kwh in the total power required for automobile manufacturing could be achieved.

Finally, it should be noted that the actual social benefit derived from an automobile is not its mere existence, but its *use*. Hence, the power required to produce an automobile is in effect amortized over the life of the vehicle (in the same sense that the initial cost of an automobile is amortized over its life). In the postwar period the durability of passenger automobiles has declined, while the durability of trucks has been maintained at a relatively high rate. We estimate that the U.S. now produces 40% too many automobiles each year, assuming that the care in production and maintenance of trucks *could* be applied to passenger automobiles.[5] Applied to the amortized power requirements for automobile metals manufacture this improvement would increase the effective power saving Modification II in terms of metals production from 56% to 72%.

It should be emphasized that the purpose of this exercise is not to determine the precise amounts of power that could be saved by modifying the design and construction of the automobile, but rather to provide a concrete demonstration of the elasticity of this power requirement and of the kinds of changes that would be involved if power-savings were undertaken. The results support the general view that, apart from the serious economic effects of such an undertaking, it would be possible to achieve considerable savings in industrial power consumption with no appreciable change in the amount or social value of the goods produced.

RESIDENTIAL POWER CONSUMPTION

One of the distinctive features of nearly all problems of environmental pollution is that the pollutants are produced by operations which benefit relatively few individuals, while the resultant social costs are broadly distributed among the population as a whole. Thus, the extremely intensive use of electric power by an aluminum plant benefits directly the relatively small number of individuals who are employees or stock-

[5]Calculations from material in AMA, 1964, 1971, and U.S. Dept. Comm. 1971a.

holders of the firm, but the resultant pollution affects the entire community. This disparity between the localization of the benefits and dispersal of the social costs associated with power production must be taken carefully into account in developing judgments that relate to human welfare.

At first glance, it would appear that this problem might be considerably minimized in the case of residential power consumption. However, as can be seen from what follows, here too there is a kind of ecological inequity, with the distribution of power consumption—and therefore of the origin of the resultant pollution—sharply polarized with respect to economic class as illustrated by the following analysis of some very recent data kindly made available for this study by the U.S. Department of Commerce (1971c, and EEI 9R-309), regarding the distribution of power-requiring equipment among households with different incomes.

This analysis deals with total resource energy consumption (rather than with electric power consumption alone) in order to facilitate the addition of another very powerful source of environmental pollution which serves households—the automobile. Figure 8 shows the distribution of aggregate energy consumption (i.e., of all the households in a given economic class) due to large appliances of convenience—which together account for about 1/3 of the total residential expenditure of

Figure 8 Estimated aggregate electric energy use for selected conveniences by income group

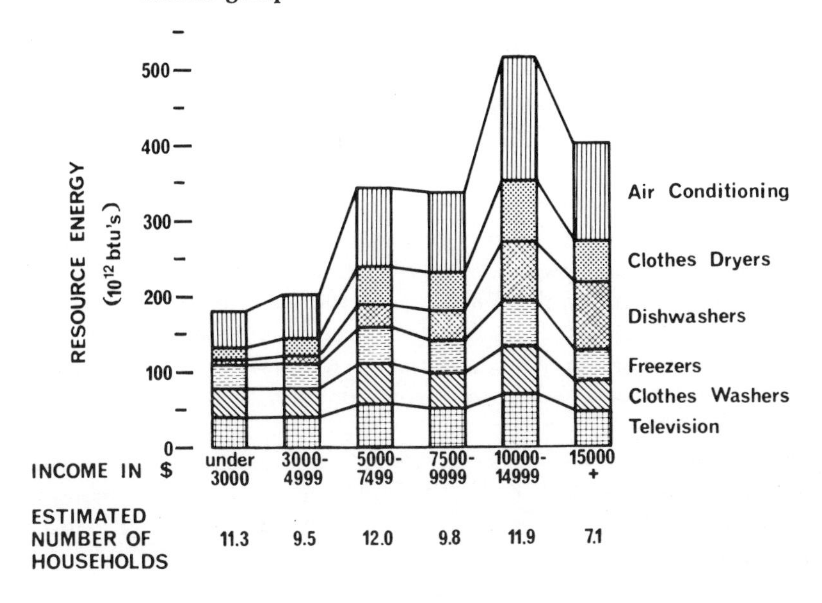

electricity—among different economic classes. The approximately 20 million households in the two wealthiest groups consume, in the aggregate, more than twice the power used by the approximately 20 million households in the two poorest groups. Between the poorest group (under $3,000 annual income) and the wealthiest (incomes of $15,000 and over) there is an approximately four-fold increase in energy consumption *per household,* the largest effects being due to air-conditioners.

An important feature of these data is their relation to the matter of "saturation" in household use of electric-powered appliances. The differences between the levels of energy use in different economic classes reflect the relative number of households within each class which operate a given appliance. Evidently, as income increases, the household's use of such utilities approaches saturation—i.e., the household uses more of the available appliances on the market. Thus computations of future levels of residential consumption of electricity can be related to the progressive saturation of all households with respect to types of appliances. This relationship is illustrated by Figure 9 which shows that the course of future residential power consumption projected by the Federal Power Commission would lead to total saturation of all households with all appliances presently available at about 2000. By these means, the residential demand for power has sometimes been linked to a socio-economic process which is supposedly underway now in the United States —upward mobility with respect to income.

What elasticity in power demand can be found in the residential sector? Table 3 summarizes a tentative effort in this direction. Two methods of achieving power savings are proposed. In those cases (water heating, space heating, cooking ranges, clothes dryers) in which non-electric appliances exist on the market, the proposed saving is achieved by the very simple expedient of eliminating the electric types. This has the added advantage of saving resource energy as well, since direct heating by local combustion of fuel, for example to produce hot water, uses about 44% less fuel than electric heating. The second method of power-saving employed in Table 3 is improved efficiency of power use. In the case of refrigerators and freezers improved efficiency is achieved at the expense of a relatively minor inconvenience—the need to defrost the appliance at intervals. Frost-free types are rather inefficient in converting power to cooling (standard refrigertors use 32% less electric power than frost-free types) because some power is used for heating in order to melt the frost.

In one important case—air-conditioners—appreciable improvement in efficiency is possible with no loss in convenience at all. The relevant evidence is summarized in Figure 10. This reports the cooling efficiencies of various models, of three typical brands, in relation to their cost (as

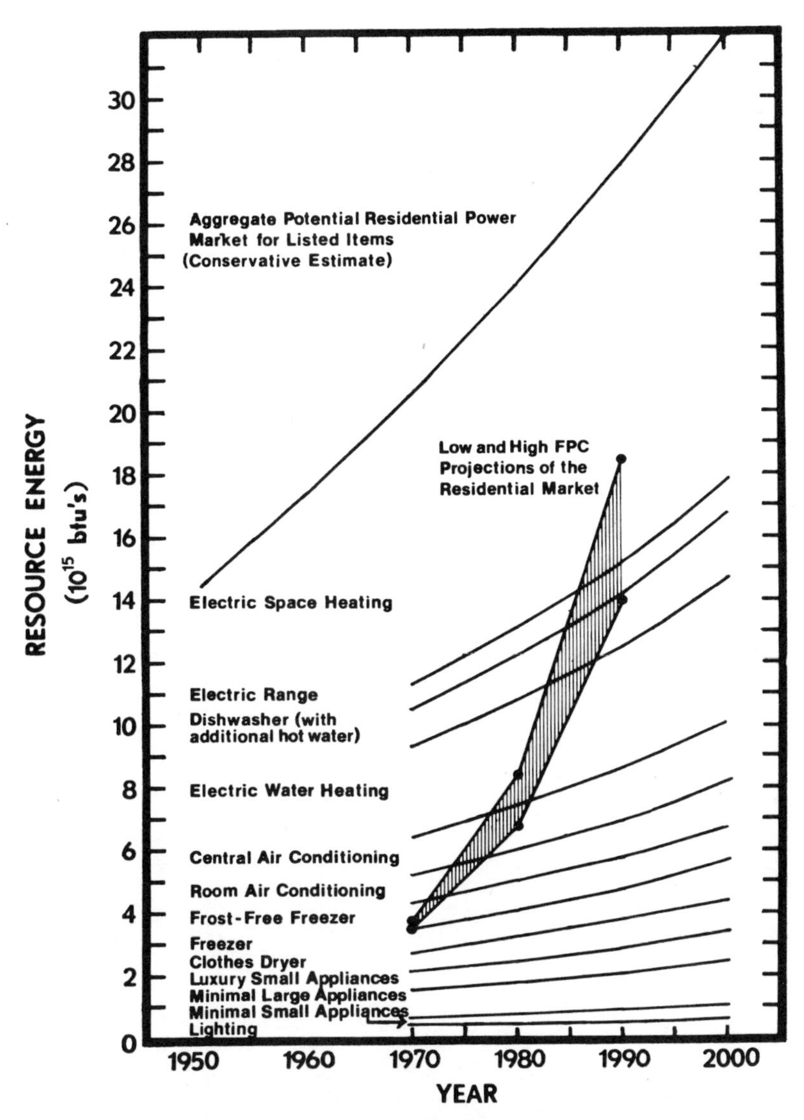

Figure 9 Potential all electric residential market
Sources: U. S. Dept. Comm., 1970, p. 35. Landsberg, p. 517. U.S. Dept. Comm. 394, p. 25, Table A, p. 25, No. 394. F.P.C. projections are from Phyllis Kline, Office of Economics, FPC.

expressed in dollars per BTU of cooling capacity). Efficiency is given as an index number which expresses the cooling achieved per unit of electricity used. Several relationships are evident in Figure 10: (a) There is a considerable overall variation in air-conditioner efficiency, from a

Table 3 Possible Power Savings in the Residential Sector

	% Households With Electric Feature[a]	1968 Aggregate Electric Consumption (in 10^{12} BTUs)[1]	Possible % Reduction	Method of Power Saving	Reduction in BTUs (Electricity)
Refrigeration[2]	99.7	250	20	Eliminate frost-free refrigeration	50[b]
Water Heater[3]	26.1	223	100	Eliminate electric heaters	223
Space Heat[4]	4.8	164	100	Eliminate electric heaters	164
Air-Conditioner[5]	36.7	154	44	Construct for maximum efficiency	68[b]
Television	99.0	128	0		0
Cooking (ranges)[6]	47.0	96	100	Eliminate electric ranges	96
Food Freezer[7]	27.2	80	16	Eliminate frost-free types	13[b]
Clothes Dryer[8]		51	100	Eliminate electric types	51
Other[9]		244	0		0
		1,390			665

[a]U.S. Dept. Comm., 1969, p. 704, unless otherwise indicated.
[b]This denotes a case where savings in energy were obtained by using a more efficient electrical unit.
[1]O.S.T.
[2]EEI-9R-309.
[3]EEI-9R-309.
[4]Moyers, AAA-R-A 79; ASHRAE 1967.
[5]Association of Home Appliance Manufacturers, 1971; Consumers Research Incorporated, 1972; and Moyers, AAA-R-A-79.
[6]EEI-9R-309.
[7]EEI-9R-309.
[8]EEI-9R-309.

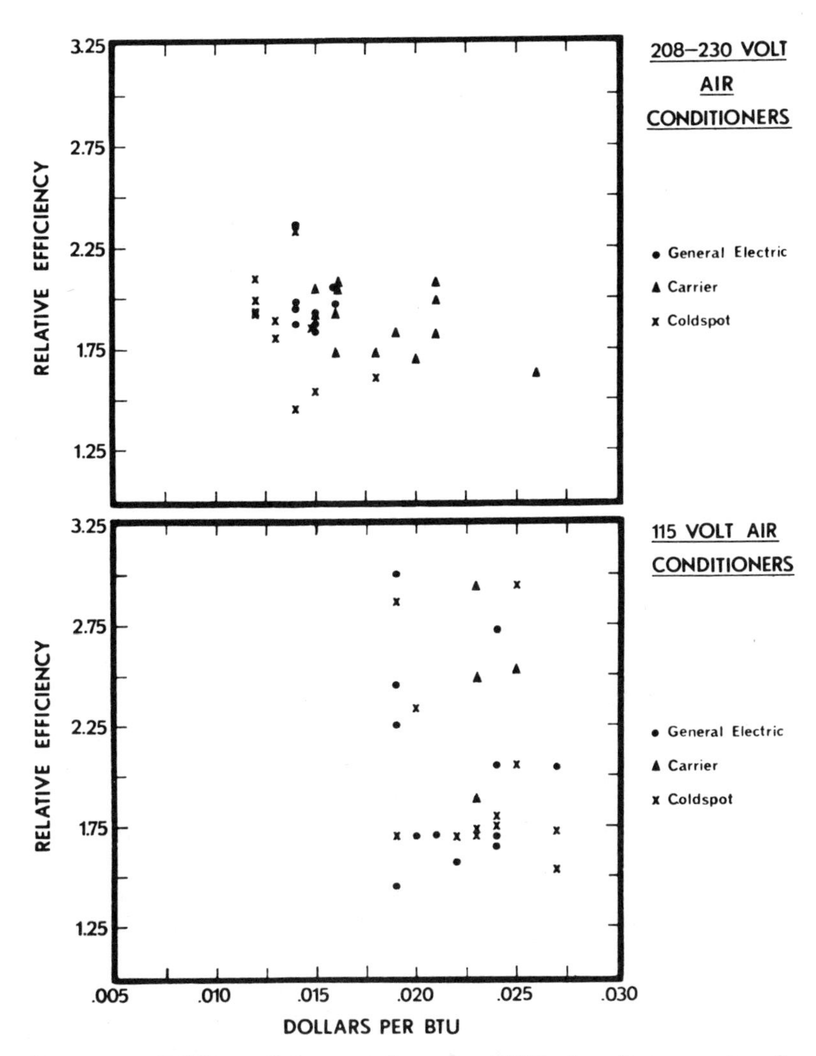

Figure 10 Relative efficiency and cost per BTU of cooling capacity of various air-conditioner models. Different symbols indicate three different brand names.

Sources of raw data: Association of Home Appliance Manufacturers, 1971 *Directory of Certified Air Conditioners*, No. 4, Chicago, Oct. 15, 1971; and Mr. Ken McFarland, Friedrick Corp., pers. comm.

minimum of about 1.5 to a maximum of 3.0. Variations are greater in 115-volt air-conditioners than in 230-volt air-conditioners; the latter are clustered in the range of about 2.0. (b) Generally, 230-volt air-conditioners cost less (per BTU of cooling) since two pole 230-volt motors are more compact (and inefficient) than the four pole 115-volt motors. In addition condenser-construction affects the efficiency of the unit. It might be noted that one recent advertisement for a central air-conditioning unit offers two models—one is advertised as maximally efficient, but presumably expensive; the other is advertised as unusually economical, but presumably less efficient.

The data of Figure 10 shows that there are important opportunities for power savings, at no cost in social value, by improving air-conditioner efficiency. Assuming that most air-conditioners previously sold were in the lower range of efficiencies, we estimate that a requirement that all air-conditioners be designed to the present maximum efficiency would result in an aggregate saving of about 36% in power consumed with no change in cooling output. Improved house insulation could probably increase this saving to about 44% (Moyers, AAA-R-A-79).

As indicated in Table 3, by these means, overall domestic power consumption could probably be cut to about half its present value. A good deal of this reduction could be achieved through increased efficiency of air-conditioning; this is a particularly important result in view of the critical effects of air-conditioning power demands on supply. It is obvious that the proposed changes would increase the domestic demand for fossil fuels, especially gas. The implications of such a shift are discussed below.

COMMERCIAL POWER CONSUMPTION

Possible elasticity in power consumption is analyzed in Chapter 3 by Richard G. Stein, who concludes that in the special case of the modern commercial sky-scraper, savings of about 50% of operating power requirements could be achieved by proper design (use of windows that open, efficient heating and air-conditioning, reduction in excessive illumination). Table 4 is a tentative effort to extend this analysis to the total commercial sector. Total possible savings of about 22% of the power used in the commercial sector, or about 4.8% of total U.S. power consumption are indicated. Note that these savings involve no loss in social value except that involved in the use of power for advertising and display lighting.

SOME ALTERNATIVE COURSES OF ACTION

Any solution to the developing power crisis in the U.S. requires, for the immediate future, either the continued exponential increase in supply or a considerable decrease in demand, and ultimately a levelling off of both. Solutions which depend upon continued exponential growth of power supply have the disadvantage of requiring equally heroic meas-

Table 4 Possible Power Savings in the Commercial Sector

End Use	Electric Power Used in 1968 10^{12} BTUs	%	Possible Savings Method	10^{12} BTUs	%	Additional Energy Needed 10^{12} BTUs
Total Input[a]	1079	100.0		260	22.0	46.2
Water Heating[a]	84	7.8	Changeover to gas[c]	84	7.8	39.5
Refrigeration[a]	244	22.6				
Air Conditioning[a]	370	34.3	10% lighting reduction[b]	37	3.4	
Cooking[a]	8	0.7	Changeover to gas[c]	8	0.7	6.7
Other[a]	373	34.6			10.1	
Lighting[b]	201	18.7	36% reduction[b]	75	7.8	
Advertising and display Lighting[b]	27	2.5	Total elimination	27	2.5	
Elevators[b]	40	3.7				
Fans and Air Handling Equipment[b]	35	3.3	Opening windows[b]	7	0.6	
Pumps and Motors[b]	18	1.7				
Miscellaneous[b]	52	4.7				

[a]OST, 1971.
[b]Stein, Richard, Chapter *18* of this volume.
[c]EEI-9R-309, *op. cit.*

ures to control environmental pollution, and eventually run up against the insuperable barrier imposed by the one pollutant which cannot be fully controlled—heat. On these grounds alone it becomes essential that we examine the alternatives which require that the current rate of growth in the demand for power be reduced and eventually brought to nearly zero.

Solutions to the power crisis which rely on reducing the demand for power present two alternative routes: demand for power can be diminished either by reducing the output of the goods and services which consume power, or by improving the efficiency with which power is used in producing goods and services (i.e., power productivity), or by a mixture of both means. Until recently it has often been assumed that pollution due to power production, like environmental pollution, generally in an industrialized country such as the United States reflects the high output of goods, and can be reduced only if such affluence is diminished. This approach assumes that the *utilization* of power—power productivity—is uniformly near optimum efficiency, so that there is little hope of reducing power input, without reducing production, by means of measures to improve efficiency, or power productivity.

Perhaps the single most important conclusion to be drawn from the data presented is that there is, in fact, considerable opportunity for improving power productivity in the United States. The chief evidence that power productivity can be improved is the fact that it has *declined* in the last 25 years: industrial power consumption has increased considerably faster than industrial output. This trend is a consequence of the technological changes which have so strikingly transformed U.S. industry since WW II: the substitution of synthetic chemical products for natural ones, the displacement of steel by concrete and aluminum, and most pervasively the substitution of machines for human labor. Each of these displacements has substituted a power-intensive process for a more power-thrifty one, so that inevitably the power consumed *per unit of goods produced* has increased.

All of the foregoing changes have been *voluntary* rather than forced by dwindling supplies of the displaced materials. The loss of steel markets to aluminum is not due to low supplies of steel. Contrary to the claims of detergent manufacturers, synthetic detergents have not displaced soap because the supplies of fat needed for soap manufacture became inadequate (the U.S. now exports enough saponifiable fat to produce soap in sufficient amounts to replace the present use of detergents). Synthetic rubber does not supplant a failing supply of natural rubber; countries like Malaysia have suffered seriously because the demand for their rubber crops—and therefore the price of rubber—has declined. Nylon cordage is not a substitute for a shrinking supply of jute cordage; the economic plight of a nation like Bangladesh, whose main export is jute, has been worsened by the shrinking *demand* for jute, as it is driven off the market by synthetics. Finally, our unemployment figures testify to the fact that automated machinery has not been introduced into industry because the labor supply has become inadequate.

I have cited these facts to emphasize a single point: that at least in material terms the industrial displacements which have so seriously intensified the demand for power are *reversible*. It becomes of interest, then, to consider what savings in power might have been effected in U.S. industry if the trends which have reduced power productivity had not taken place.

As noted earlier, overall power productivity of U.S. industry has declined from 0.68 1958 dollars in 1947 to 0.44 1958 dollars in 1967, a period in which production, as measured by value added increased considerably. If this increasing production, after 1947, had been carried out at the overall power productivity of 1947, a 35% savings of power could have been effected at no sacrifice of production (as measured by value added). This can be extended to an estimated 38.6% savings in industrial power in 1970.

By corresponding means one can compute the effect on power consumption of preventing the displacements, after 1947, of various power-thrifty processes with power-intensive ones. Thus, for example, if the 1947 ratio of steel to aluminum *production* had been maintained, power consumption by the metals industries would have increased much more slowly than it actually did—to 22% less than its actual 1967 value.

Finally, one can of course reduce industrial power consumption simply by eliminating the production of goods which are regarded as socially undesirable. Thus, the elimination of ordnance production would have reduced the 1967 industrial power consumption by as much as an additional 0.8%. Along the same lines, we estimate that 4.25% of the power sold in 1970 had its end use in activities of the Department of Defense. Approximately 50% of that was used directly by the Defense Department and the other half was used in production of various goods for Defense Department consumption. Hence, a reduction in defense activities would effect a reduction up to 4.25% in the nation's demand for power.[6]

That power productivity has also declined sharply in the commercial sector has been amply demonstrated by Richard G. Stein and others.

[6]This estimate (from *Scientific American* 1971) is a rather gross one. For a fuller consideration of the actual impact of the defense establishment on the U.S. energy budget, other parameters would have to be taken into account to determine to what extent energy currently used by the DOD, such as that which goes into the maintenance of troops, etc., would have to be expended whether or not there was an army. Further consideration of the problems attendant upon the demilitarization of the U.S. economy are to be found in Leontief, 1961.

Technological changes in building materials and design have not only increased the power cost of erecting buildings, but have sharply increased the power required to operate buildings. Lights are operated continually, at unnecessarily high intensities of illumination. Sealed buildings are unable to take advantage of outside air to regulate internal temperatures; with increasing frequency buildings and even single rooms are heated and cooled concurrently. These transformations, too, are reversible, and one can estimate from data representative of the entire commercial sector, that such changes could have reduced 1970 commercial electric power consumption by up to 22%. Adding this to the savings in industrial power suggested above, we arrive at a total reduction in 1970 power consumption of about 22% in these two sectors.

It is appropriate at this point to take note of some of the economic consequences of the foregoing means of reducing industrial power consumption. Most of these are self-evident. Constraints on the substitution of aluminum for steel and lumber would mean that the use of aluminum would need to be restricted to products in which it contributes an essential and non-trivial function, for example aircraft. In contrast, aluminum furniture might be eliminated in favor of wooden types. Constraints on the rapid expansion of the chemical industry would mean that cotton and woolen fabrics would predominate over synthetic fibers; plastics would be used, not as substitutes for paper or wood (as in containers and furniture), but only where their unique properties are essential (as in shatterproof windows).

Apart from these reversals in the displacements among industrial products, attention must be given to the most significant displacement which has accompanied the rapid growth of industrial power consumption in the United States—the displacement of labor by electricity. The inverse relationship between labor productivity and power productivity has already been alluded to, and it has been pointed out that increases in power productivity such as those proposed above, probably require a disproportionate increase in labor. Obviously, some of this increase would involve the re-introduction of hand labor in place of machine operations—for example, as a result of the re-introduction of wooden furniture in place of the molded plastic variety. Similarly, if the chemical, cement, and aluminum industries were restricted in output there would have to be more jobs filled as their products were displaced by the products of less power-consumptive (and therefore, usually more labor-intensive) industries and agriculture.

Here we arrive at perhaps the most basic change that would be involved in a course of action which proposed to relieve the power crisis by reducing demand. In the industrial sector, labor productivity—i.e., value added/man-hour—would necessarily decline. In the U.S. economic system, this is a catastrophic prospect, as is evident from the serious concern created by the recent reduction in the *rate of increase* of labor productivity. This concern is apparently warranted, for the profit yielded by the economic system is strongly dependent on labor productivity—and apparently also on a continually rising labor productivity. On these grounds, there would appear to be a fundamental clash between the requirements of the private enterprise system and any effort to improve the power productivity of industrial production, and thereby reduce power consumption and the attendant environmental degradation. This conclusion parallels one reached earlier with reference to pollution problems generally. (See Commoner, 1971.)

Finally, what can be said about the possibility of reducing the rate of growth of the residential demand for power? We pointed out earlier that the issue here is whether low income groups are to be prevented from achieving the high power expenditure level characterized by the wealthy households which are nearly saturated with available appliances, or whether the appliance-saturated households are to be deprived of power. (This question along with the implications of variation in life style for power demand are discussed in Chapter 22. That treatment raises the potential of further reduction of household power demand by selective use of house appliances.) However, certain savings of residential power seem to be possible without sacrifice of convenience and services. This is especially true of electric space and hot water heating which represents an increasing share of residential power consumption and is notably inefficient in terms of resource energy, as compared with direct combustion of fuel for heat.

Since about 40% of present residential electric power is used for space and hot water heating a considerable reduction in the present average amount of electricity used per household could be achieved by avoiding the use of electricity for these purposes. Clearly such a move would increase the demand for residential gas fuel, which is in increasingly short supply. A great deal of gas is consumed by the synthetic chemical industry; since this is a highly power-consumptive industry, and since natural substitutes are available for many of its products, it might be reasonable to divert gas from the industry to residential use. Although such a move would exacerbate the economic problems already involved in the industrial changes discussed earlier, it seems reasonable, then, to adopt the resultant savings in residential power consumption, which, along with the other savings in Table 3 account for 47.8% of residential power and about 13.8% of total power—as an approximation of what might be achieved in this sector.

As computed for 1970, a total saving in electric power consumption of about 35.8% would appear to be possible (Table 5). Given the considerable approximations involved in these computations, this value

Table 5 Possible Savings of Electricity Consumption in 1970

Sector	Billion KWH Used in 1970	Percent of Total U.S.	Savings			
			Method	Billion KWH	Percent of Sector	Percent of Total U.S. Consumption
Industrial						
sales	577		If operated at 1947 rate of power productivity (Fig. 4)	257	38.6	17.4
generated less sold	89	45				
sub total	666					
Commercial	308	21	Table 4	68	22.0	4.6
Residential	448	30	Table 3	214	47.8	13.8
All other sales	58	4			0	0
Total U.S.	1,480	100		539		35.8

Sources: U.S. Statistical Abstracts 1971, p. 501.
Bureau of the Census, 1967 Cens. Mnfg. *Fuels & Electric Energy Consumed* p SR4-11 (for ratio of industrial power purchased to industrial power generated less sold.) OST, op. cit. for ratio of commercial to industrial power consumption.

ought to be regarded only as an indication of the order of magnitude of possible power savings that might result from the indicated changes in use-pattern.

Our intent in the foregoing discussion has been to explore some of the ways in which patterns of power utilization would have to be changed in order to confront the power crisis by reducing the projected rate of growth in demand. What are the consequences of this exercise?

First, even on the basis of the fairly limited analysis which we have carried out thus far it would appear that power demand is quite responsive to variations in the pattern of use, and that appreciable savings could probably be obtained by such a strategy. This is evident in the 35.8% reduction in present power consumption that might have been achieved by the series of shifts in use-pattern, which were outlined previously. On these grounds, the overall strategy has merit.

However, in one fundamental sense, the strategy of reducing power demand by regulating the pattern of use has a portentous consequence: it probably cannot be achieved without sweeping changes in the economic factors which apparently govern industrial production, and the distribution of wealth. Clearly the types of changes in patterns of power use just described would result in widespread dislocation of industrial production. Considering the intensity of the protests over the loss of several thousand jobs incurred by abandoning the SST program, how difficult would it be to restrain the growth of aluminum production and the construction of concrete roads in favor of wider use of steel and lumber and railroads? And beyond such difficulties lies the even graver matter of the apparently unavoidable dilemma created by an effort to reduce the overall power demanded by industrial production: Either total production is curtailed, or power productivity is elevated; but whichever course is taken, the effort to reduce power demand would appear to clash head-on with one or both of the two factors that are widely regarded as essential to the stability of the U.S. economic system—increased production and increased labor productivity.

These considerations raise the possibility—which it is to be hoped economists will investigate—that the continued exponential increase in power consumption is not an accidental concommitant of industrial growth, but is rather a functional necessity for the continued operation of the U.S. economic system, as it is presently organized. If this should prove to be true, then the ultimate social choice signified by the power crisis becomes very stark. One course is to continue the present exponential growth in the supply of electric power, and risk our future on the ability to contain the huge mass of resultant chemical, radioactive, and thermal pollution. The other course is to slow down the rate of power consumption, and accept as a necessary consequence that the economic system must be changed.

It is sometimes said that the current concern with environmental problems is a "motherhood issue"—a benign problem which provides a convenient way to divert attention from serious social inequities such as poverty and unemployment. The relationship between power production—which is a major segment of the environmental crisis—and human welfare suggests the reverse. Any effort to find a solution to the power crisis is certain to engage, at the deepest level, the nation's concept of social justice.

Is it just that a wealthy household should contribute so much more heavily to power consumption—and therefore to the resultant pollution—than an impoverished one, when the environment is the common property of both? Is it just to offer an aluminum chair for sale without warning the buyer that its production—in place of a wooden one—has worsened environmental pollution? Is it just that wages of labor are based on its productivity (a practice apparently now to be enforced by national policy) when any effort to improve the efficiency of industrial use of power—and thereby relieve its stress on the environment—is likely to reduce that productivity?

The power crisis, like every other environmental issues, is not an escape from the responsibilities of social justice. It is, rather, a new way of perceiving them.

Acknowledgments

The authors would like to thank Paul Stamler and Dru and George Lipsitz of the AAAS Committee Staff and LaVerne Papian and Howard Boksenbaum of the Center for the Biology of Natural Systems for their suggestions and assistance in the preparation of this chapter. The authors are particularly indebted to Caleb A. Smith for his critical review and technical advice with respect to our computations regarding the productivity of labor, resource energy, and power in the manufacturing industries.

REFERENCES

American Society of Heating, Refrigerating and Air Conditioning Engineers, 1967. *Guide and data book equipment.* New York.

Automobile Manufacturers Association, 1964. *Auto facts and figures.* New York.

———, 1971. *Auto facts and figures.* New York.

Bravord, J. C., et al., 1971. *Energy expenditure associated with the production and recycle of metals.* ORNL.

Cambel, A. B., et al., 1965. *Energy R & D and national progress.* U.S. Government Printing Office, Washington, D.C.

City of New York Environmental Protection Administration, 1971. *Toward a rational power policy.* Apr.

Commoner, Barry, 1971. *The closing circle.* New York: Knopf, Chap. 12.

Consumer Research Inc., 1972. Air Conditioning, Fans. *Consumer bulletin annual 1972.* Washington, N.J.

Edison Electric Institute (EEI). *Appliance comparison reference of electric energy consumption with fuel use.* EEI, N.Y., EEI-9R-309.

———, 1970. *Pocketbook of electric utility industry statistics, 16th edition.* EEI, N.Y.

———, 1971. *Questions and answers about the electric utility industry.* EDI, N.Y.

Ethyl Corporation, 1952. Brief passenger car data for 1952. N.Y.

Federal Power Commission. *An economic model for residential electricity.* Washington, D.C.

Landsberg, Hans H., et al., 1963. *Resources in America's future.* Baltimore: John Hopkins Press.

Leontief, Wassily W., and Marvin Hoffenberg, 1961. The economic effects of disarmament. *Sci. Am.* 204, no. 4.

Lyon, William H., and D. S. Colby, 1951. *Production consumption and use of fuels and electric energy in the United States in 1929, 1939, and 1947.* Report of Investigation 4805, U.S. Department of the Interior, Bureau of Mines, U.S. Government Printing Office, Washington, D.C., Oct. (I 28.23:4507)

Makhijani, A. B., and A. J. Lichtenberg, n.d. *An assessment of energy and materials utilization in the U.S.A.* Paper presented at the Department of EECS, U.C. Berkeley, Siminar on the Ecology of Power Production.

Moyers, John C., *The value of thermal insulation in residential construction.* ORNL-NSF EP9 NSF I.A. No. AAA-R-A-79.

Office of Science and Technology. *Patterns of energy consumption in the United States,* unpublished.

Pennoch, Jean L., 1964. The household service life of durable goods. *J. Home Economics* Jan.

Risser, Hubert E., 1970. *Environmental Geology Notes.* no. 40. Illinois State Geological Survey, Urbana, Ill. Dec.

The Editors of Scientific American, 1970. The input/output structure of the United States economy, chart.

Seattle City Light Co. *Operating costs of your electric servants.* Pamphlet 3-67-50M, Seattle, 1967.

U.S. Dept. of Commerce, Bureau of the Census. *Current population reports,* No. 394.

———, 1949. *Census of manufactures, 1947.* U.S. Government Printing Office, Washington, D.C.

———, 1957. *Census of manufactures, 1954.* U.S. Government Printing Office, Washington, D.C.
———, 1961. *Census of manufactures, 1958.* U.S. Government Printing Office, Washington, D.C.
———, 1966. *Census of manufactures, 1963.* U.S. Government Printing Office, Washington, D.C.
———, 1969. *Statistical abstract of the United States, 1969.* 90th ed. Washington, D.C.
———, 1971a. *Census of manufactures, 1967.* U.S. Government Printing Office, Washington, D.C. (including special Report SR4, *Fuels and energy consumed*).
———, 1971b. *Statistical abstract of the United States, 1971.* 92nd ed. Washington, D.C.
———, 1971c. *Survey of consumer buying expectations,* unpublished data.

BRUCE HANNON

Bottles, Cans, and Energy Use[1]

In an energy shortage, the most efficient energy systems will dominate, but we must also soon question the need of the quantity consumed in providing a given quality of life. In other words, there is an extremely important distinction between evaluating system efficiency and evaluating system energy flow; the former is primarily an engineering matter, the latter is mainly a societal one.

Had energy not been so accessible and therefore so cheap in the U.S., many systems might not be so extensively in use today, e.g., the automobile, the aluminum industry, the packaging industry, etc. But energy is rapidly becoming expensive. Dr. David Schwartz, assistant chief of the Office of Economics of the Federal Power Commission, estimates that the internalization of environmental costs alone will raise the cost of electricity about one cent per kilowatt hour which should more than double the current industrial prices. We therefore must plan for least-energy-consuming systems. In fact, some hold that energy is the real currency in systems made of living components (Odum, 1970). Under this view, items would have as their true value the resource energy commited in their manufacturing. This type of pricing reflects wasted energy and dwindling energy resources, as well as the scarcity of the mass involved.

THE PROBLEM

The packaging industry has blossomed in the last few years. The American consumer paid about $25 billion in 1966 for packaging, ninety

[1]Reprinted, in part, from *Environment*, March 1972, with permission. I wish to thank George Voss, John Hrivnak, and James Benton for their efforts in this systems analysis.

percent of which was discarded (EPA, 1969). By 1971 the packaging industry had become an even larger multibillion dollar interest, opposing the reduction of the amount of materials used in wrapping, sacking, canning, bottling and otherwise protecting and selling merchandise to the consuming public. Based on convenience appeal to our increasingly mobile society, advertising campaigns, principally by container manufacturers, have convinced the public that the throwaway container works to their best interest. But litter, problems in solid waste disposal, high consumer purchase cost, and resource drain are fostering a worthy adversary to the packaging syndrome.

Nowhere is this conflict more clearly underscored than in the packaging of beer and soft drinks. These two commodities alone constitute a major portion of the food and beverage consumed by the U.S. public and about one-half of all beverage and food containers (EPA, 1969).

The first efforts at market conversion to one-way beverage containers were made by the steel industry in the late forties and early fifties. Together with the major can companies, they viewed the beer and soft drink market as the last major expansion area for steel cans. With returnable bottles averaging about 40 trips from the consumer to the bottler, it was clear that 40 cans would be needed to replace each returnable bottle over a period of six to eight months. Aluminum companies made their entrance into the market in the mid- to late-fifties by introducing the all-aluminum beer can. Aluminum has since appeared in the tops of steel beer and soft drink cans to facilitate opening.

A surprising early success of cans was found in the ghetto areas of major cities. Because inner-city dwellers generally travel on foot and have small family storage and cooling space, they often purchase one cooled package per visit to the market; the three to four cent deposit per single returnable bottle lacks the appeal of a case of returnable bottles. Further, the process of reacquiring the deposit appears to be very demeaning for the inner-city resident (Joyce Bottling Co., 1970–71). There was also stiff retailer resistance to accepting returned bottles because of diminishing retail storage space. In 1960, for example, 40% of the roofed supermarket space was devoted to nonselling storage space, and in 1970 only 10% of such space was for storage. However, the bottler's storage space has increased significantly due to the diversity of containers (Trebellas, 1971).

Suburbanites tend to favor returnable bottles much more than do inner-city residents. Suburbanites usually do their grocery shopping by auto, and also can exert more influence over the retailer. As a result, returnable bottles make significantly more trips back to the grocer in the suburbs.

Since the beverage makers and bottlers were not selling containers to the public, they were indifferent to the needs of the steel and can companies. Nevertheless, the beverage wholesalers and retailers had their own inducements toward conversion of the market to throwaway containers; the reduction of storage space and the elimination of the labor of sorting and stacking returnables had obvious appeal.

At about the same time, glass bottle manufacturers realized the impact of cans on their market, and competition between glass and steel throwaways began. With these pressures from both sides, the bottlers revised their bottling lines to handle throwaway containers.

Therefore, the decline in return rate and retreat from the market place of the returnable is not caused by bottle fragility but is due to general affluence, competition from other packages, and advertising, the mechanism for change in consumer habits.

The container manufacturer uses advertising not only on the consumer but on the other key parts of the industry. The Glass Container Manufacturers Institute's advertising program was described by an official of that organization as "three-pronged, directed at the packer, the retailer, and the ultimate consumer."

In the fact of this campaign, the market share of the returnable bottle has declined appreciably and further declines are expected. While beverage consumption rose 1.6 times from 1958 to 1970, beverage container consumption rose 4.2 times during the same period, a further reflection of the increasing use of throwaway containers. While fewer returnable containers are being sold, those which are used make fewer return trips to the bottler. From a high of 40 return trips per soft drink container in the early fifties, for example, the national average has declined recently to about 15 trips.

There is still reluctance among some bottlers to continue the throwaway practice because the one-way container increases the cost of consumption and therefore reduces sales and consequently profit. Many claim, however, that the throwaway container has increased their sales volume and accordingly improved their competitive position.

But the bottlers are succumbing to the "economies of scale" and following the centralizing tendencies of other industries. For example, there were 8000 soft drink bottlers in 1960; a decade later their numbers had dwindled to only 3600 (Joyce Bottling Co., 1970–71). The centralized bottlers tend to the one-way container because of the inefficient return shipping of empty returnable bottles.

The only force opposing this centralizing tendency is the franchising procedure used by the major beverage makers. These franchises provide each bottler an exclusive territory, a practice which tends to irritate the large food wholesalers who are then required to buy soft drinks locally. This procedure breaks up the food wholesalers' economy of scale and has consequently produced the "private-label" soft drink. These brands are packaged exclusively in cans and sold over extremely large areas from very large centralized plants. The private-label soft drinks are sometimes sold with the wholesalers' or retailers' label and tend to be priced much lower than the major brands. The price difference is in the beverage, however, not the container.

REMELTING AND RECYCLING

A plethora of bottle and carton sizes has accumulated on the market. Beverages are sold in 7-, 10-, 12-, 16-, and 28-oz. containers and sold singly, in 6-packs, in 8-packs or in 24-bottle cases. This great variety tends to conceal the unit cost from all but the most calculating consumer. For example, six 12-oz. soft drink returnables sell for $1.04, including a 30¢ deposit, while six 12-oz. cans sell for 99¢. The marked price encourages can sales, but the beverage in returnable bottles is actually about 30% less expensive than in cans or throwaways. A recent court ruling in Illinois declared it illegal to tax the consumer on the deposit charge. This has had the effect of requiring a clear display of the deposit charge.

A 12-oz. returnable soft drink bottle costs the bottlers about 9¢; a 12-oz. throwaway bottle costs about 4¢ and a 12-oz. can costs about 5¢ (Joyce, 1971). Even though the glass throwaway costs less than a can, it is heavier and subject to inconvenient breakage. Cans can be shipped more compactly than one-way glass bottles and are much less difficult to dispose of. Throwaway glass containers can be returned to the bottle maker as waste glass (cullet), although color separation is a considerable problem. The small percentage of returned cans are not remelted but chemically dissolved to obtain copper, which is currently very scarce in natural form.

Besides the decrease in number of trips made by the returnable bottle, other results of heavy emphasis on throwaway packaging were a sharp increase in roadside litter and a significant increase in solid waste. In particular, those who become committed to the one-way container were quite concerned with the adverse advertising provided by a labeled discarded container. The agencies that had formed to promote their throwaway container interests now joined together to form an anti-litter organization. The Can Manufacturers Institute and the Glass Container Manufacturers Institute teamed with the U.S. Brewers Association and the National Soft Drink Bottlers Association to form the vehicle for public education against littering, "Keep America Beautiful, Inc." It was through advertising campaigns by these agencies that I became interested in the packaging industry. It initially seemed a paradox that these same agencies would vigorously and successfully oppose the reduction in generation of solid waste and litter and at the same time promote anti-littering campaigns.

In answer to the solid waste problems, the above agencies have actively supported a solid waste collection system which would gather household and commercial waste and separate the waste into recyclable components from which new products could be made. This proposed solution closes, with the exception of lost materials, the mass flow loop in packaging beer and soft drinks. The industry calls this "remelting" concept "recycling," and it is. But returnable bottles are also a form of "recycling," perhaps best called "refilling." However, the system in use today seems to be tending toward exclusive use of the one-way container without the collection, sorting, and remelting system. Thus, we have three systems, two of which claim to be recycling and all of which satisfy the same consumer demand with respect to beverage consumption.

Since the economic differences between the systems go relatively unnoticed and the recyling of mass is available for each system, we need some better device for underscoring the system differences. No more suitable means exists to my knowledge than to examine the total resource energy required to operate these systems for a given quanity of beverage, and energy-effectiveness analysis.

THE ENERGY ANALYSIS

An energy analysis has been completed on the entire soft drink, beer, and milk container systems. Bottles, cans, paper, and plastic containers were considered. The analysis is here described in detail only for the soft drink industry and only the energy associated with the containers, not the beverage, has been tabulated.

The analysis is a comparison of two systems—that which delivers a given quantity of soft drinks in throwaway containers, and that which delivers the same quantity in returnable glass containers. Each system is evaluated with and without a remelting loop for discarded containers; the containers may be ultimately thrown away and/or collected and remelted. The energy required to operate each system for one gallon of beverage is calculated and the results compared.

I have neglected the indirect energy commitments in the processes mainly because they seem small and difficult to calculate. Such energies would be those required to make the bottling machines, delivery trucks, paper and plastic packaging of the metal and glass containers, etc. As an example of the order of magnitude of such energies, the energy required to build an automobile is about 5% of the energy consumed in gasoline in the car's lifetime. Likewise, the energy of human labor is neglected as negligibly small. Rough calculations by the author indicate that this indirect energy is about 0.5% of the fossil mechanical energy expended in the container industry. This figure is supported by M. Tribus and E. C. McIrvine (1971), who calculate the ratio at 0.4%. The total energy consumed by the industry, including such things as building, heating, and lighting, is given in this study.

The energy calculations are based on the quantity of basic energy resources removed from the ground. Thus, a kilowatt-hour of electricity is taken as 11,620 British Thermal Units (BTUs) since the coal-, oil-, or gas-fired steam-electric plants are only about 29.4% efficient (1968) in converting and delivering the resource energy to the user, and these are the major (99%) electricity sources in the Midwest (EEI, 1969). The amount of electricity generated by nuclear plants is quite small relative to coal, oil, and gas, and can be neglected (Hubbert, 1969; Gast, 1971). Since mining and transportation of coal to a power plant take about 0.25% of the mined coal energy, we assume this factor is also small for petroleum, and the resource energies of a gallon of gasoline and of diesel fuel were taken at their stated values of 125,000 BTUs per gallon and 138,000 BTUs per gallon respectively (Bureau of Census, 1968; Bureau of Mines, 1969). The natural gas net energy value used was 1000 BTUs per cubic foot. Propane has an energy content of 88,888 BTUs per gallon at 30 inches of mercury and 60°F. (National Tank Co., 1969).

The resource energy required to ship one ton-mile of freight by rail was taken as 640 BTUs (ICC, 1969). An inner-city diesel truck consumes 2400 BTUs per ton-mile of cargo (ICC, 1969; Rice, 1940).

The number of trips per returnable bottle is of key importance in the analysis. According to the Illinois bottling industry (Joyce, 1971), the average return rate in the state is about ten; in rural areas the average is about thirteen, while the Chicago average is eight.

A study of a downstate Illinois small-city bottler revealed a return rate of 24. Much of his returnable bottle sales were through gasoline service stations, milk delivery routes, and drink dispensing machines.

The bottler who supplied the data for this study sells over 70 million soft drinks annually. His all-urban data is so extensive that it allowed an accurate comparison between the 16-oz. returnable, the 16-oz. throwaway bottle, and the 12-oz. can. These are the major sale items, with several million fillings annually in each form. The average number of refills for the 16-oz. returnable is eight (Joyce Bottling Co., 1970–71). The average unit weight of the 16-oz. returnable bottle is

Table 1 Energy Expended (in BTUs) for One Gallon of Soft Drink in 16-Ounce Returnable or Throwaway Bottles

Operation	Returnable (8 Fills)	Throwaway
Raw material acquisition	990	5,195
Transportation of raw materials	124	650
Manufacture of container	7,738	40,624
Manufacture of cap (crown)	1,935	1,935
Transportation to bottler	361	1,895
Bottling	6,100	6,100
Transportation to retailer	1,880	1,235
Retailer and consumer	—	—
Waste collection	89	468
Separation, sorting, return for processing, 30% recycle	1,102	5,782
Total energy expended in BTUs per gallon:		
Recycled	19,970	62,035
Not recycled	19,220	58,100

1 pound and 0.656 pound for the 16-oz. throwaway bottle. Thus, for each gallon of beverage which flows through the soft drink industry, one 16-oz. returnable or eight 16-oz. throwaway bottles are required. These bottle groups would weigh 0.005 tons and 0.002625 tons respectively. (See Table 1.)

The returnable bottle materials are acquired from the raw materials and transported to the glass manufacturer, who blends new materials with recycled glass to make new bottles which are transported to the bottler. The returnable bottle makes a number of cycles through the outlet and consumer and back to the filler before breakage or loss requires its transport to the collected waste system. Here the household and commercial wastes are separated from their glass content at about 50% efficiency. Further sorting and cleaning has 60% efficiency, and thus 30% of the glass in the received waste is returned to the glass manufacturer for reprocessing. The nonsalvageable glass goes on to other uses or to a landfill disposal site. The separation, cleaning, and sorting operations are presently in pilot-plant stage, but the container manufacturers insist that this is the method of the future. The Bureau of Mines of the U.S. Department of the Interior has a process which removes glass from the ash residue of a solid waste incinerator.

The throwaway bottle has a similar history except that the entire flow goes through the collection, separation, cleaning, and sorting process.

A small percentage of the throwaway containers is also currently being disposed of in glass collection centers and sent to the container manufacturers for reprocessing.

Table 1 summarizes the results of my calculations of the energy resources needed for each step in the manufacturing, distribution, and return or disposal of glass soft drink containers. The numbers are given in terms of energy per gallon of beverage, and are stated in BTUs per gallon. The estimates are based on Chicago-area statistics, where a returnable bottle makes an average of eight return trips. These data are later scaled to reflect fifteen fillings, which is the national average (Table 3).

As noted earlier, I have calculated energy requirements in terms of the energy value of the resource—coal, oil, or natural gas—required for each step of the container system. Where electrical energy is used, I have calculated the fuel needed to produce the energy used.

The energy costs of material acquisition are the costs of mining sand, limestone, soda ash, and feldspar, the raw materials for glass manufacture, in the quantities and proportions needed (Glass Industry, 1971) to manufacture one 16-oz. returnable bottle, which weighs 1 pound and in eight return trips would carry a gallon of beverage, and to manufacture eight throwaway 16-oz. bottles which weigh 5.25 pounds. Transportation energy for carrying these raw materials to the manufacturer is based on statistics, provided by one of the nation's largest glass container manufacturers, which indicate that the average distance between the manufacturing site and the source of raw materials is 245 miles, and that 79% of these materials was moved by rail and 21% by truck. In the process of container manufacturing, 15 to 20% of the newly made glass comes from internally recycled glass, or cullet, the scraps generated in the manufacturing process itself (this, by the way, is the basis for the statement occasionally heard that 20% of the glass in this industry is already being recycled). According to the container manufacturer already cited, there would be no significant increase in energy costs to increase the proportion of cullet from internal (bottle manufacturers) or external (consumer) sources. The total energies of the container industry were derived from a detailed examination of overall industry data (Bureau of Census, 1971). The cost of cap or crown manufacturing was calculated separately (CPA, 1969; Bureau of Census, 1971, 1971a).

From the Census of Transportation (CPA, 1969; Bureau of Census, 1971, 1971a) it was found that the average distance between the bottle manufacturer and the bottler was 345 miles, and that 70.2% of the empty bottles was shipped by truck, 16.3% by rail. (The remainder was shipped by trucks belonging to the container industry; therefore, this energy has already been included.) These data were in turn used in calculating the energy of transportation to the bottler, with an added correction for the fact that, according to the inner-city trucking companies, a single truck capacity averages 1500 cases of throwaways but only 1200 cases of returnable bottles (Stout, 1971; McVey, 1971).

Heating and lighting account for a large share of the energy cost of bottling—filling the empty bottles with soft drink. Because of this there are no economies of scale in energy costs of bottling; the major urban bottler used in this study was compared to a small local bottler with one-tenth the sales, and it was found that the major bottler used 6.3% more resource energy per gallon of beverage than the small firm.

Eight gallons of water are needed to wash each returnable bottle before refilling, but only about one-half gallon of water is needed to wash a freshly manufactured throwaway (Joyce, 1971). This would lead to a higher energy cost for returnables in this regard, but the major bottler used in this study is required by municipal law to wash returnables and throwaways in the same manner. From our study of the small bottler, we estimate that about 7% of the throwaway bottling energy could be saved if the less extensive rinse cycle were employed. Warehousing is included in the total bottling figure.

Transportation to the bottler was weighted to reflect the more space-consuming returnable bottles, which were assigned 55% of the allocated energy for storing and moving. According to the Census of Transportation, the average trip from bottler to retailer is 231 miles, 74% of which is by private truck.

No energy charge has been made for the retailer and consumer; soft drinks account for less than 1% of supermarket space, and storage and labor costs are small compared both to the total for the containers and to other demands on the retailer. No effort has been made to estimate the energy costs of the return of reusable bottles by the consumer, since it is assumed that only rarely would a trip to the supermarket or grocery store be made solely for the purpose of returning empty bottles.

The costs of collection of discarded bottles are based on information provided by local sanitary haulers (McLaughlin, 1971), who say that the average truck capacity is four tons, that trucks have a 50% load factor and average seven miles to a gallon of gasoline.

It should be noted that can and glass collection centers, which are now a major interest of some environmental groups, are economically infeasible (EPA, 1969). Can and bottle manufacturing companies subsidize these intense activities, which in any case require donated labor and free transportation. (Paper collection and recycling is, fortunately, economically feasible.)

Rather than concluding this analysis with the costs of disposal of discarded or broken bottles, I have calculated the additional costs of returning this glass to the bottle manufacturer—what is most often called "recycling." This is done in two steps: glass must be separated from other trash in the municipal waste system, and then decontaminated and sorted. Only one firm is now making the equipment to perform these tasks. The unit is called the Hydro-sposal and is capable of reclaiming paper, glass, metal, textiles, and plastics. Pilot-plant operations (150 tons per day) have begun in Franklin, Ohio (Black Clawson Co., 1971), and it is data from this plant I have used for my calculation.

Half the glass is lost in the separation from trash, and only 60% of the remainder is recovered from the sorting and decontamination steps. Hence, 70% of the discarded bottles, both returnable and throwaway, must be disposed of into a nearby landfill. The cost of sorting and separating all of the glass is assigned to the 30% which is returned for remanufacture. No energy charge is made to landfill material.

The transportation energy to return scrap glass back to the manufacturer is estimated as the transportation energy of the raw materials which it would replace.

It is important to note that the energy cost of retrieving scrap glass from waste is far higher than the energy cost of mining raw materials, and that therefore, from an energy standpoint, it makes little sense to recycle, in the sense of remelt, either returnable or throwaway bottles. The only energy savings are realized in reuse of the container; retrieving it as raw material is wasteful of energy. Substantially more (sixty times) energy is used to obtain waste glass as a replacement for crushed stone in making road-pavement materials.

Without reviewing the calculations in detail, I have tabulated a similar analysis for the soft drink can, which is given in Table 2. These data were derived principally from the Bureau of Census records (Bureau of Censes, 1968, 1971, 1971a).

The energy estimates for steel and aluminum are for "finished shapes and forms" manufactured from ore; the estimates consequently do not distinguish between structural steel and tin plate sheet steel. About 30% of the finished steel came from scrap, while no aluminum scrap is included in the calculations.

Adding the energy costs in the various categories in Tables 1 and 2, we find that returnable bottles are far superior from an energy standpoint to throwaways, either bottles or cans. From Table 1 we can see that remelting scrap bottles, while returning about 30% of the glass, would consume more energy than manufacturing all of the glass from raw materials. Removing this item from the total, therefore, and assuming that all bottles eventually find their way into landfills, we see

Table 2 Energy Expended (in BTUs) for One Gallon of Soft Drink in 12-Ounce Cans

Operation	BTUs per Gallon
Mining (2.5 lbs. of ore per lb. of finished steel) (*Packaging Digest*, 1971)	1,570
Transportation of ore (1,000 miles by barge) (Rice, 1970)	560
Manufacture of finished steel from ore (Bureau of Census, 1971)	27,600
Aluminum lid (11.9% of total can weight 4.7 times the unit steel energy) (Bureau of Census, 1971)	12,040
Transportation of finished steel 392 miles (Rice, 1970)	230
Manufacture of cans (4% waste) (Bureau of Census, 1971)	3,040
Transportation to bottler (300 miles average) (Rice, 1970)	190
Transportation to retailer	6,400
Retailer and consumer	—
Waste collection (McLaughlin, et al.)	110
Total energy for can container system[a]	51,830
Total energy for 12 oz. returnable glass system	17,820
Ratio of total energy expended by can container system to that expended by 12 oz. returnable glass:	2.91

[a]The all-aluminum can system consumes 33% more energy than the bimetal (steel and aluminum) can system.

that the energy cost of returnable bottles is 19,220 BTUs per gallon of beverage, assuming each bottle makes eight return trips, the case in Chicago. This compares with 58,1000 BTUs per gallon for throw-away bottles, and 51,830 BTUs per gallon for cans. In other words, throwaway bottles use 3.11 times the energy of returnable bottles, and cans use 2.70 times the energy of returnables.

If we extrapolate these data to the nation as a whole, we assume that each returnable bottle is used fifteen times, the present national average, instead of eight. In this case, the energy cost of each gallon of soft drink in returnable bottles would drop to about 13,150 BTUs per gallon, and the throwaway bottle and can would be still more expensive by comparison. Throwaway bottles would then require 4.4 times the energy of returnables (both 16-oz.), and 12-oz. cans would use 2.9 times the energy of 12-oz. returnable bottles.

Although soft drink containers consumed nationwide slightly less than 0.17% of the nation's total energy in 1970 (beer containers consumed the same amount) (Hubbert, 1969; Abrahamas, 1970), much can be learned from analyses such as this one. It is possible to show that a complete conversion to returnable bottles would reduce the demand for energy in the beverage (beer and soft drink) industry by 55%, without raising the price of soft drinks to the consumer (indeed, returnables are cheaper than throwaways). A complete turn to throwaways, of course, would substantially increase the energy demand of beverage containers.

Finally, we should point out that dollar costs do not reflect energy, a result of the cheapness of energy as compared to other inputs in manufacturing (although these costs are increasing). A ratio of the cost of the returnable 16-oz. glass bottle system, in dollars per gallon of beverage, to the cost of the throwaway system was found to be 2.1 (Trebellas, 1971). To compare consumer costs the beverage costs must be included. Thus, with an average of fifteen returns for each returnable, throwaways cost about 1.2 times as much as returnables to make and distribute in Illinois. As noted earlier, the actual ratio of prices is now 1.3 as an Illinois average, which indicates that a significantly higher profit is being made on throwaways than on returnables. H. Folk indicates that a complete return to returnables in Illinois would save consumers of the state $71 million annually (Folk, 1971).

Similar energy calculations have been made for beer bottles and cans, and for plastic, paper, and glass milk containers. The results are generally similar to the case of soft drinks. The energy ratios (that is,

Table 3 Energy Ratios for Various Beverage Container Systems (The Energy per Unit Beverage Expended by a Throwaway Container System Divided by the Energy per Unit Beverage Expended by a Returnable Container System)[a]

Container Type					
Throwaway	Returnable	Quantity	Beverage	Returnable Fills	Energy Ratios
Glass	Glass	16 oz.	Soft Drink	15	4.4
Can[b]	Glass	12 oz.	Soft Drink	15	2.9
Glass	Glass	12 oz.	Beer	19	3.4
Can[b]	Glass	12 oz.	Beer	19	3.8
Paper	Glass	1/2 gal.	Milk	33	1.8
Plastic[c]	Plastic	1/2 gal.	Milk	50	2.4

[a]Without remelting loop (discarded bottles and cans are not returned for remanufacture).

[b]Bimetallic-aluminum and steel.

[c]High-density polyethylene.

the energy expended by a throwaway container system divided by the energy expended by a returnable container system) are presented in Table 3.

In short, from dollar-cost and energy-cost standpoints, returnable bottles are preferable to cans or throwaway bottles. Can systems require approximately the same energy as throwaway bottles. Throwaways provide convenience to the consumer and additional profits to manufacturers and retailers. The question of which system is preferable to society as a whole cannot be resolved until both sides of the balance are known, and until there is some means of securing the system most desirable to society as a whole, rather than the one most profitable to manufacturers and retailers.

REFERENCES

Abrahamas, J. H., 1970. Utilization of waste container glass. *Waste Age* July–Aug.: 72.

Black Clawson Co., Middleton, Ohio, 1971. Letter to Bruce Hannon, Mar. 17.

Bureau of Census, 1968. 1963 *census of mineral industries*, vol, 1. U.S. Dept. of Commerce, Washington, D.C.

———, 1971. 1967 census of manufacture. U.S. Dept. of Census, Washington, D.C.

———, 1971a. 1967 census of transportation. U.S. Dept. of Commerce, Washington, D.C.

Bureau of Mines, 1969. IC 8411 and IC 8401. U.S. Dept. of Interior, Washington, D.C.

Bureau of Solid Waste Management, 1969. *The role of packaging in solid waste management*. Environmental Protection Agency, Washington, D.C.

Folk, Hugh, 1972. Employment effects of the mandatory deposit regulation. Illinois Institute for Environmental Quality, 189 W. Madison St., Chicago. Jan.

Edison Electrical Institute, 1969. *Edison Electrical Institute Statistical Handbook*. New York.

Gast, Peter, 1971. Atomic Energy Commission, Argonne National Laboratory, Letter to Bruce Hannon, June 23, 1971.

The Glass Industry Jan. (1971): 31.

Hubbert, M. King, 1969. Energy resources. Chapter 8 of *Resources and man*. National Academy of Sciences, New York.

Interstate Commerce Commission, 1970. Transportation statistics in the U.S. for 1969. Washington, D.C.

The Joyce Bottling Company, 1970-71 data. 4544 W. Carroll Ave., Chicago.

Joyce, Tom, 1971. Solid waste—Litter. Release to State of Illinois Subcommittee on Solid Waste Reduction, The Joyce Bottling Company, Chicago, Aug. 10, 1971.

McLaughlin and Sons, Cumins & Westmore, Sanitary Haulers, Urbana, Illinois, 1971.

McVey Trucking, Oakwood, Illinois, 1971.

National Tank Company Handbook, 1959. National Tank Company, New York.

Odum, Howard, 1970. *Environment, power, and society.* New York: John Wiley & Sons.

Rice, R. A., 1970. *System energy as a factor in considering future transportation.* American Society of Mechanical Engineers, New York, 70 WA/ENER-8, Nov. 1970.

Stout Trucking Company, Urbana, Illinois, 1971.

MICHAEL CORR AND DAN MACLEOD

Home Energy Consumption as a Function of Life-Style

We build power plants, tolerating air, water, and thermal pollution as well as massive alterations of the landscape, in order to create products and comforts for ourselves. Exactly which products and comforts our power resources are used to produce are determined by our life-style, defined here as the manner in which resources are used to fulfill human needs. Life-style, in turn, can be looked at as being dependent on family structure, values, and so on, and as these factors vary, so does life-style. Viewed in this manner, life-style can yield some interesting insights into home energy consumption. Conversely, the conditions of resource availability to families or cultures may affect their life-style. Since energy may become more expensive (less available), it is of interest to examine life-styles which are less dependent on energy for what intuitive guidance they might give us in our thoughts about the elasticity of demand for energy in the residential market.

The historical trend and recent policy preferences have been to small families and, as the number of large household operations has decreased, many of the advantages of large families have been lost. For instance, in 1900, with an average of 4.8 people in the household, chores like dishwashing, laundry, gardening, and canning could be shared among several people; now, with an average household of 3.2 people, the dishwasher, the automatic laundry, and the freezer seem like necessities to the housewife.

Nuclear families, consisting of mother, father, and children, and extended families, which include grandparents or other relatives as well, are not, however, the only kinds of large households (with size-related advantages) possible. During the past few decades, considerable interest in various patterns of communal living has been exhibited by young people. Students, artists, young working people, political activists, and religious people have formed numerous large households with varying degrees of success (Snyder, 1969). A true commune is defined as a household with a common pool of income and property, allocated by means of a single budget; a common home, in which all goods and facilities (for example, kitchen and bathroom) are fully accessible to all members, with consensually determined provisions for privacy; and a common provision for the care and education of children. A life-style that leads to communal use of facilities would appear to make a pronounced difference in personal energy consumption.

Unfortunately, only meager data to support an analysis of life-style as a function of family structure are available through the Bureau of Census of the Department of Commerce. The bureau data only indicate that major appliance ownership, and, therefore, presumably domestic energy consumption, increases with family income. In addition, thorough analysis of communal use of resources relative to use by other types of households is complicated by the wide variety of life-style innovations among urban and rural communes.

In an attempt to gain insight into commune resource use, a group of twelve communes, with a total of 116 members, in the Minneapolis area were visited, and a questionnaire was administered to determine patterns of natural gas, electric power, and motor fuel consumption, as well as appliance and automobile ownership.

The limitations of this study are many. The houses were selected rather arbitrarily through personal acquaintances and may not be a representative sample of the entire communal counter-culture. Minneapolis has a relatively severe winter requiring more energy use for heating. Most groups did not keep accurate records. Utility bills were sometimes paid with the rent. Odometers in some communes' automobiles were broken. Thus, some figures must be considered rough estimates. Nevertheless, the study does give some interesting indications of the degrees of reduction in energy consumption which could result if large numbers of people began sharing consumer goods and otherwise simplifying their life-styles. For example, computed on a per household member basis, natural gas, on the average, was consumed by the commune members at a rate 40% below that for an average Minneapolis house of 900 square feet occupied by 2.6 people, electric power at a rate 82% below the Minneapolis average, and gasoline at a rate 36% below the national average.

Natural gas consumption by all of these groups was similar, whereas power use varied widely from group to group, and the consumption of gasoline varied even more widely (in some cases being even higher than the national average). This results perhaps from the fact that the

Table 1 Summary of Direct Commune Energy Use

	Natural Gas (Minneapolis)	Electricity (Minneapolis)	Gasoline (Nationwide)
Average per household	120.6 thousand cu. ft./yr.[a]	6,651 kwh/yr.[b]	905 gallons/yr.[c]
Average per person	3,860 cu. ft./mo.	213 kwh/mo.	285 gallons/yr.
Commune use per person			
low	1,700 cu. ft./mo.	17 kwh/mo.	60 gallons/yr.
average	2,300 cu. ft./mo.	39 kwh/mo.	181 gallons/yr.
high	3,000 cu. ft./mo.	66 kwh/mo.	444 gallons/yr.
Percentage reduced per person			
low	56%	92%	79%
average	40%	82%	36%

[a]Based on estimated average housing unit size of 900 sq. ft. A. L. Pooler, Minneapolis Gas Co., and Allan Anderson, Minnesota State Planning Agency.

[b]Personal communication, Margaret Brian, Northern States Power, Minneapolis; and U.S. Dept. of Commerce, Bureau of Census, *1970 Census of Housing, Minneapolis-St. Paul Urbanized Area, Block Statistics,* Aug. 1971; figure for average household of 2.6 members.

[c]*U.S. Statistical Abstracts,* 1971, p. 36, 3.17 people per household. U.S. Bureau of Consumer Buying Expectations, unpublished data, 1.291 vehicles per household. *U.S. Statistical Abstracts,* 1971, 700 gallons of motor fuel consumption in 1969 per automobile year.

consumption of power and gasoline is dependent on personal habits as well as the size of the household (Table 1).

The low rate of energy consumption was voluntary and was not considered by the people involved as a lowering of their standard of living. Most of the people in these groups are young political activists who decided to live collectively for social and financial reasons. The reduction in energy consumption was a result of this desire to live collectively and to de-emphasize the materialistic aspects of life. Some of the people were making conscious attempts to reduce their consumption of energy; others were not. Some were apologetic about their high bills and the number of appliances they owned; others had not given energy conservation much thought.

APPLIANCES

In an attempt to quantify the difference in appliance use between communes and other households, appliance ownership was computed on an average per capita basis for the commune sample and compared with the averages for all households in the United States. Some striking differences were observed. Whereas about 25% of Minneapolis homes had air-conditioning (Brian, 1972), the communes had none. The absence of air-conditioners is particularly significant, since air-conditioning, in addition to using great quantites of power (around 1500 kwh per room unit year), adds greatly to the summer power demand peaks which often cause our brown-outs. The communes had no dishwashers or food freezers and only two clothes dryers for the 116 members, 0.02 dryers per person, or only one-eighth the national average. Some commune members may be hanging their clothes on clotheslines to dry, but most are probably using dryers at the laundromat where they wash their clothes.

Washing clothes at a laundromat may not save any "laundering energy," but as a communal facility, laundromat machines will typically do the work of several home laundry devices, thereby saving considerable secondary energy, that is, manufacturing and materials extraction. Obviously, it takes more energy to produce washers for several individual families than it does to produce one laundromat machine, which can serve the same number of families. For their home laundry, air drying clothes and ironing only essential items could have saved additional energy for commune members. Grandma's method of using a wringer washer to wash whites, then darks in the same hot water, would use only 25% as much hot water as an automatic washer doing the same laundry (New, 1966; Seattle City Light).

It is difficult to estimate the possible impact of communal living on energy required for manufacturing. It seems likely that appliances in large households are used more intensively than those in average-sized households. There is more food to cool, cook, or blend; more work for toasters, irons, and record players. Commune appliances would therefore seem more likely to wear out, rather than be retired due to obsolescence. (This does not necessarily hold true for such appliances as refrigerators, which run all day regardless of how many people use them.) When consumers use articles until they wear out, instead of discarding them for other reasons such as obsolescence, they help relieve the energy burden of manufacturing.

In an attempt to evaluate the demands of the "new life-style" on the manufacturing sector, saturations were computed from the commune data for nineteen appliances or conveniences. Household saturation for a convenience is the percentage of all households which own or are furnished with that convenience. Because the average commune had ten members, more than three times the national average for households, appliance saturations were computed on a per capita, as well as a household, basis. These data are listed in Table 2.

The national average household saturation for the nineteen conveniences considered was 74%, and the commune average household saturation was 62%, somewhat lower, certainly, but not strikingly so. What was striking was the difference among per capita saturations. The national average was 24%, the commune average only 6%.

Note that every one of the conveniences considered, from cars to blenders, can serve a communal function. It can be argued that as long as a household has a convenience, all the members enjoy its services. From this general observation, one can see that the commune members are being served much better by their investment in appliances than are average occupants. Their per capita *access* to conveniences is almost equal to that of the average U.S. household member even though their per capita *consumption* is only 25% of the national average, measured in terms of saturation. Thus, such a life-style clearly consumes much less secondary energy through the nation's manufacturing sector than does the average household, on a per capita basis.

When the convenience saturations were weighted by the approximate new retail value of the conveniences involved, the pattern apparent in the saturation averages was reinforced, in spite of the wide range of prices between toasters and cars. The average commune had $6715.30 worth of the conveniences considered, 31% more that the $5116 computed from national household saturations for the conveniences considered. The average new value per commune individual was $671.53, which was 59% below our estimated national average per household individual. In terms of energy required for producing, promoting, and retailing such conveniences, input-output table calculations indicate that the average U.S. household member has indirectly used 123.9 million British Thermal Units (BTU) of energy through the nineteen conveniences considered. The average commune member has used only 49.97 million BTUs, or 59.6% less (*Scientific American*, 1970; Hirst, 1972). These energy savings can be amortized over an estimated ten-year appliance lifetime, yielding a saving of about 7 million BTUs per person each year for commune individuals. It can be estimated that U.S. energy consumption per capita was about 351 million BTUs per capita in 1971. Therefore, by saving 7 million BTUs (after amortization) the commune member is achieving a 2% saving in per capita energy use.

The energy economies in actual use of the conveniences turn out to be comparable in magnitude. As noted earlier, the average commune member used 40% less natural gas and 82% less electricity than the

Table 2 Selected Convenience Saturation for New Life-Style Communal Groups Compared to the National Average

	National Average				Sample Communes		
Convenience[a]	Approx. Price (if new)[b]	No. per Household[c]	No. per Individual	Weighted Price per Individual	No. per Commune	No. per Commune Member	Weighted New Price per Commune Member
Cars and Trucks	$2,910 retail			$1,192			$533
	$2,195 wholesale	1.291[d]	0.41[e]	900	1.84	0.19	418
Air-conditioner	250	0.25[f]	0.09	23.20	—	—	—
Dishwasher	200	0.21[d]	0.07	14.00	—	—	—
Freezer	225	0.33[d]	0.10	22.50	—	—	—
Television	175	1.22[d]	0.38	66.50	0.68	0.07	12.23
Refrigerator	300	0.998[g]	0.32	96.00	1.36	0.14	42.00
Radio	50	0.998[g]	0.32	16.00	1.65	0.17	8.50
Electric iron	15	0.997[g]	0.32	4.80	0.75	0.08	1.20
Toaster	12	0.93[g]	0.29	3.48	0.87	0.09	1.08
Water heater	100	0.92[h]	0.29	29.00	1.07	0.11	11.00
Vacuum cleaner	75	0.92[g]	0.29	21.75	0.68	0.07	5.25
Coffee maker	14	0.89[g]	0.28	3.92	0.19	0.02	0.28
Mixer	25	0.82[g]	0.26	6.50	0.29	0.03	0.75
Washing machine	175	0.74[d]	0.23	40.25	0.17	0.02	3.50
Range	275	0.59[h]	0.19	52.30	0.97	0.10	27.50
Frypan	20	0.56[d]	0.18	3.60	0.2	0.03	0.60
Electric blanket	16	0.50[g]	0.16	2.56	0.39	0.04	0.64

Clothes dryer	150	0.47[d]	0.15	22.50	0.17	0.02	3.00
Blender	25	0.37[g]	0.12	3.00	0.39	0.04	1.00
Average saturations		0.74	0.24		0.62	0.06	

	National	Commune
Weighted new retail value per person		
Without autos and trucks	$ 431.86	$ 118.53
With autos and trucks	1,623.86	671.53
Weighted new retail value per household		
Without autos and trucks	$1,356.40	$1,185.30
With autos and trucks	5,116.40	6,715.30

[a]Since national household saturation data were available for only 19 of the 32 conveniences, such items as hairdryers, waffle irons, electric tools, and one mimeograph machine were omitted from the calculations.

[b]Approximate prices for all appliances are from *Sears and Roebuck's Fall and Winter 1971 Catalogue*, except for the air-conditioning price, which was taken from their 1972 Spring and Summer catalogue.

[c]Except for ranges and water heaters, all saturations are for 1971.

[d]U.S. Bureau of Census, Survey of Consumer Buying Expectations, July 1971, unpublished data.

[e]Assumes 3.17 people per household, the national average for 1970; U.S. Dept. of Commerce, 1971, op cit.

[f]Saturation of air-conditioners for Minneapolis only; personal communication from an undisclosed source at Northern States Power.

[g]U.S. Dept. of Commerce, *Statistical Abstracts*, 1971, Table 1117.

[h]U.S. Office of Science and Technology, *Patterns of Energy Consumption in the United States*, household saturations for 1968.

Table 3 Average Per Capita Energy Savings Tabulated for New Life-Style Participants

Savings Category	Usage Reduction (in millions of British Thermal Units)	Savings as Percent of National per Capita Energy Consumption[a]
Motor fuel	13.1	3.7
Gas, oil, and electricity	47.7	13.6
Amortized energy for manufacture and distribution of household conveniences (over ten-year period)	7.0	2.0
Totals	67.8	19.3

[a]Base national per capita energy consumption: 351 million British Thermal Units in 1971.

average household individual. This yielded 71% energy saving over average per capita residential consumption.[1] Residences account for 19.2% of end use of energy (OST, 1971); thus, the commune members achieved an additional saving of 13.6% of national average per capita energy consumption through their economies in direct natural gas and electricity use (assuming 1968 proportions of energy use for 1971). Moreover, the commune members used 36% less motor fuel than the per capita average of 285 gallons per household member, a saving of 104 gallons. This is a saving of 13.09 million BTUs per person, or a saving of 3.7% of national per capita energy consumption. Altogether, the *savings* of energy by commune individuals amount to 19.3% of total U.S. national per capita energy consumption. (See Table 3.)

Since consumption of household goods (new value) for commune members was 60% below the national average, it can be surmised that the cash needs of commune members were considerably less than the national average for adults. Many communes have special businesses, such as running a bakery, a repair shop, a printing shop, a radio station, a newspaper, a free school, a Yoga school, or a documentary film operation. Communes also save energy in such cases where home

[1]In 1968, the residential sector satisfied 34.8% of its requirements by petroleum, 50.1% by gas, and 15.1% by electricity according to the Office of Science and Technology. However, if energy losses of 72% during generation of electricity are included, these figures for direct home energy dependence become 25% for heating oil, 36% for gas, and 39% for electricity. Multiplied by the commune saving in each case, this yields a total home energy saving of 71% ($1.0 \times 0.25 + 0.40 \times 0.36 \times 0.82 \times 0.39 = 0.25 + 0.14 + 0.32 = 0.71$).

and work are the same place for the members. For instance, one bakery and restaurant commune manages to get by with only two small buses for seventeen adult members and two children.

TRANSPORTATION

In the commune sample there were 32 vehicles, and the average fuel consumed per vehicle was 655 gallons per year, 6% below the national average. Motor fuel consumption per individual was 36% below the national average, and the automobile mileage per individual was 68% below the national average. However, the average commune automobile was driven about the same mileage as the average domestically owned U.S. automobile. By ride sharing, economies were possible in going to school, work, and for shopping and other errand running. Mass transit use by commune members was minimal; only four individuals of the 116 members commuted by bus, a method of passenger travel with about one-fourth the impact on energy resources per passenger mile as auto travel (Luszczynski, 1972). The twelve communes also had 45 bicycles, more than one for every three people. Bicycles use about 4% as much energy as automobiles per passenger mile (Hirst, 1972a).

In 1968, the total energy used to produce and distribute new cars in the U.S. was 2040 trillion BTUs (Hirst, 1972a). This amounted to 3.8% of all energy available for end use in the economy. The automobile's large share of the national energy budget as well as its relative importance in the family and commune budget, suggest that it is worth considering whether communal auto use patterns offer any advantages beyond the lower per person mileages observed in the communes.

Another approach to insuring that social capital invested in the auto is used in a manner which minimizes impact on the environment would be to look for ways in which the individual auto could be used more intensively. If Americans drove the same number of passenger miles by auto but did it in fewer autos (that is, more passengers per auto on the average), then the individual auto would be producing more social good per unit of auto production. That higher total mileages are possible can be seen from the experience of the taxicab industry. Taxi companies often get 150,000 miles of service before an engine needs a major overhaul (Kraus, 1972). The taxi is usually given an engine overhaul a couple of times, and total mileage in the neighborhood of 300,000 miles is not uncommon. By contrast, the average family drives 10,000 miles per auto each year (Lansing and Hendricks, 1967). In order to put the taxicab's 300,000 miles on their auto a family would have to drive it, on the average, 30 years, but in 1970 the average age of the auto on the road was only 5.6 years (AMA, 1971). Thus, when the energy required for production is considered, it appears that present patterns of family auto usage are resource wasteful, since most autos are taken out of service due to accidents or psychological obsolescence, rather than, as in the case of the 300,000-mile taxi, due to being worn out (SPWC, 1970). The fact that the autos of the commune sample were receiving slightly less than average mileage, even though the average commune had three times as many members as the average U.S. household, suggests that optimal use of the auto as commune social capital might be affected through sharing autos between communes, so that they receive intensive use, and are at least not retired due to psychological obsolescence. In Washington, the May Valley Coop (about six households) and the Ploughshare Collective (three households) have experimented with inter-household, shared vehicle ownership. In the latter case, two vehicles serve seventeen adults, two children, and the bakery and restaurant which support the collective. The Ploughshare vehicle mileage per individual is about 40% below the national average, while the vehicles are receiving about 60% more use than the average family auto (Shakow, 1972).

Commune ownership of autos (for functions which cannot be met by bicycle or bus) also offers the possibility of organized community maintenance of vehicles. Another option which would extend the useful life (total mileage) of autos would be the selection of mechanically simple vehicles for community use. This would encourage community members to perform a larger proportion of the maintenance and repair of community vehicles, a policy already followed by the Ploughshare Collective.

Clearly these notes on communal use of vehicles are not intended as the basis for curing the complex environmental problems of U.S. transportation. From Chapter 2 on transportation by Lusczcynski and Grimmer, it is clear that mass transport offers great reductions in personal energy consumption where population densities are sufficient to support it (Luszczynski, 1972). Such work also indicates that there may be great gaps in the services which can be provided by mass transit in the U.S., and that therefore additional energy savings might by necessity be required to come from changes in the cultural pattern of use of the automobile.

Examination of the transportation problem suggests that it is deeply related to cultural and technical factors. Evaluations of energy consumption through the materials used in autos and comparison of auto durability relative to that of trucks and buses, suggests a change in the design of automobiles could yield an auto which could provide more miles of use with less investment in energy during the vehicle production process (see Chapter 5, Commoner and Corr, 1971). However, as indicated in the case of taxis, social factors dictate against giving full use to a vehicle because of the psychological obsolescence factor, and the fact that the individual family has no chance to drive an auto as intensively as would be optimal. The idea is to use fewer machines more intensively, to minimize production while still meeting transportation needs. Large communes with few vehicles offer this advantage of intensive vehicle use.

A small example of cultural adroitness of implementing life-style alterations in transportation was highlighted in a recent news release from Poland. For the past ten years, at a small fee, Polish authorities have provided hitchhikers with travel coupons which are lottery tickets for the drivers who pick them up. The hitchhikers' fees support an insurance system protecting both parties in a hitchhiking-related accident, and the coupon book which is used to flag motorists assures them that everything is on the up-and-up (Walker, 1972). While in Poland hitchhiking was first encouraged because of an auto shortage, in the U.S. it may someday be encouraged to alleviate the problem of traffic jams due to excessive reliance on personal transportation.

When many of the young talk about the implications of life-style, their interest goes far beyond what happens in the commune kitchen to encompass questions of prevalent culture and economic structure. Paul and Percival Goodman's book, *Communitas*, begins to deal with these cultural questions on the national scale, considering life-style as a variable (Goodman).

HOME HEATING AND LIFE-STYLE

As a life-style problem, the transportation question immediately invites one's inquiry to the quality of urban planning and production in the automobile industry, the largest industry in the nation. By contrast, questions regarding home heating are tightly coupled to personal values and are subject to individual decisions.

Home water and space heating and cooking accounted for 90% of domestic energy consumption in 1968, exclusive of transportation (OST, 1971). Of the twelve communes surveyed, only two possessed electric space heaters. In both cases, the heaters received little or no use, and heating was done primarily by gas. Minneapolis housing units have an average of 2.6 occupants (Bureau of the Census, 1971), and in our comparison of gas consumption, the commune units were compared to 900-square-foot homes assumed to be occupied by 2.6 people, using approximate data supplied by the Minneapolis Gas Company. Thus, the homes used in our comparison averaged 346 square feet per inhabitant, while the large dwellings used by the communes averaged

2340 square feet, or 242 square feet per inhabitant. The commune members had adequate, though slightly less, space per individual than Minneapolis families living in average-sized dwelling units, while using considerably less gas and electricty for all purposes including space heating (40% less gas and 82% less electricity).

In this comparison, it appears the saving for space heat is due both to more intensive use of space, and to space heating economies of scale offered by large dwellings, with their favorable surface to volume ratio. It should be noted that multiple dwellings also offer the heating advantage, provided space utilization is achieved to as full a degree as in the case of the communes. In space heating, as in the cases of automobile use and appliance use, the communes seem to be enjoying per capita advantages of scale which are not available to the nuclear household living in their own single family structure.

Since interest in Oriental and American Indian philosophies is often high among new life-style groups, it is appropriate to examine the question of space heat through the eyes of those cultures to see if there are adaptations which might help a new life-style group reduce their energy consumption.

The cultures of Tibet and the American Indian offer an extreme in the attitude towards cold. Some Tibetan Yoguns believe that they can endure extremes of cold and heat through chanting religious verses, a practice involving breath control, and hence, perhaps, control over metabolism (Evans-Wentz, 1970). The Tlinget Indians of the Northwest American coast report in their myths having used cold salt water baths regularly to increase their endurance (Garfield et al., 1961). The Zen meditation halls of lowland Japan are usually unheated, reflecting Buddhist attitudes of indifference towards "pain." Cultural factors, no doubt, temper the attitudes of the people of India towards heat. They are reported to tolerate ten more degrees of heat at comparable humidity than Americans, before considering themselves uncomfortable (Banerji, 1959).

In correspondence with the shared Tibetan and American Indian practice, or discipline, of functioning in the extremes of the elements, one author points out that work under conditions of heat or cold can be moderately adapted to through repeated short exposures and the use of appropriate clothing, but the conditions may severely limit output. For instance, a man can do five times as much work at 90°F and 60% humidity as at 120°F and 60% humidity, and work in heat or cold can easily result in debilitating heat prostration or, in the case of cold, frostbite sometimes resulting in gangrene and loss of limbs (Basu, 1959).

The widely accepted technical approach to thermal comfort involves the determination of what spatially uniform combination of temperature, humidity, and air movement is necessary for a man to work comfortably with a given degree of clothing. For instance, one author calculates that a room with relative air velocity zero and relative humidity 50%, a lightly clothed man doing light work would require an ambient air temperature of 75°F, whereas the same man doing light work in heavy clothing would require an air temperature of 60°F (Fanger, 1970). Such criteria for optimum comfort have increased steadily by about 10°F since 1900, a *cultural change* attributed to the year-round use of lighter weight clothing, changing diet, and new comfort expectations (ASHRAE, 1972). Similarly, a decision to use twice as much energy to achieve air conditioning for weather above 70°F rather than above 75°F in New York City reflects a cultural judgment as well as the characteristics of human physiology.

Following the preference for uniform space heating noted earlier, central space heating is the prevalent approach in the U.S. at this time. One can see intuitively that this approach must require much more energy than localized heating methods, but few data are available comparing the respective energy costs. Obviously, climate places severe limitations on the application of localized heating in homes.

Aside from intuitive advantages of the English system of local heating by small fireplaces rather than space heating, a 1920s heating engineer reported that the English also accustom themselves to a room temperature of 60°F (Trane, 1923). Such a practice could yield considerable energy savings in the U.S. in regions with a climate similar to that of England, that is, the East and West Coasts.

Much of the American South and West coast is subtropic, as is Japan, a culture in which the art of localized heating has been perfected to a surprising degree. The nucleus of the traditional Japanese home energy system, which was common until well after WW II, consisted of a wood-fired bath, a charcoal-fired cooking fire, charcoal hibachis and low tables with charcoal heaters under them. The hibachi may be built into a low wooden chest, or it may be a crockery cylinder from 5 to 25 gallons in volume. It is usually lined with ashes, and the charcoal fire is built underneath a brazier which holds an iron kettle for hot water. During the coldest portion of the year, the hibachi may be supplemented with a kotatsu, or low table with a quilt bib to hold in the heat produced by a well protected tray of hot coals beneath the table.

Four Japanese-American women, who matured in early post-war Tokyo, estimate that their households used from two to four tewara-bags of charcoal per month in the winter, plus a kindling ration for the bath. Rough calculations indicate that their subsequent energy bill for heating and cooking was from one-tenth to one-fifth that of an American apartment in a city with comparable, subtropic winter (Corr and MacLeod, 1972).

As an extreme example of Japanese efficiency, it was customary for Tokyo folk to sleep with two or three quilts using a hibachi to keep the chill off the room. A few coals (one ounce) of charcoal from a hibachi will keep an insulated metal bedwarmer (placed between the blankets) warm for a full night, consuming 4% as much energy as its American counterpart, the electric blanket. The Japanese informants claimed that throughout WW II they had no difficulty at all in obtaining ample supplies of charcoal, the surplus of which was perhaps contributed to by the saturation bombings of Japanese cities. Since WW II, small gas heaters, gas heated baths, instant water heaters, and electric hibachis, etc., have come into common use, preserving to only a certain extent the traditional pattern of localized heating. Thus, our comparison of Japan and the U.S. must be considered historical with Japanese reliance on fossil fuels for home energy consumption growing rapidly (Bureau of Statistics, 1968).

Typical to Japanese culture, the sophistication of their adaptation to winter goes far beyond conveniently designed "spot heaters." For instance, in contrast to Western costume, several kimonos can be worn on top of each other, providing great warmth. When this method is applied in moderation, there is no undue loss of comfort, since the sleeves, collars, and waists, aside from being elegant, fit together conveniently.

HOT WATER, CULTURE, AND THE BATH

In addition to the energy consumed in space heating, the heating of water accounted for another 15% of U.S. domestic energy consumption in 1968, an increase of 2% from the 1960 level. Analysts usually focus on the method of heating water when discussing the increased energy bill for hot water, since an electric water heater uses about twice as much energy per gallon of water heated as a gas heater.

One expert estimates the bath might account for about 30% of domestic hot water use (New, 1966), or 5% of total domestic energy consumption. Table 4 demonstrates the important roles of both cultural and technical factors in this aspect of domestic energy consumption. As would be expected, a big, electrically-heated tub bath uses twice as much energy as the same bath heated by gas. A more energy-economical, conventional bath is a gas-heated, small tub bath using about 22% as much energy as the big, electrically-heated bath.

The large Japanese style iron tub bath heated directly by a wood fire provides the luxury of the large American tub bath at 22% of the energy cost per person. In the Japanese case, however, there is a heavy cultural cost attached: the loss of privacy. The Japanese achieve their economical bath by an ingenious system of sharing the bath water. The members of an extended family all bathe in the same tub of water, some scrubbing with soap outside the tub while each person takes a turn soaking (with no scrubbing) in the tub. Since no scrubbing is done inside the tub, the water is clean enough after five or more people have bathed to still wash a tub of dark clothes in the water. Using the conventional electric hot water heater and automatic washer, a bathing and laundry function which costs the Japanese 86,300 BTUs of energy, would cost an American household 470,000 BTUs, or about 5 times as much. Using gas technology, it would cost the American household 207,700 BTUs, or 2.4 times as much. The strength of the Japanese adaptation to bathing rests on the cultural factor of willingness to tolerate family nudity, and the technical factor of a directly heated bath enclosed in a small room which is also heated by the fire directly under the tub. Undoubtedly the fact that hot water was obtained for laundry as a by-product of the bath contributes to the low energy consumption by the traditional Japanese system.

A large Finnish-style wood-fired sauna studies in the Sierra Mountains of California achieved a level of efficiency similar to that of the Japanese system when shared by about ten people. The common bath or sauna would be acceptable in many communes since many members have a more relaxed attitude toward nudity. Under such special cultural circumstances where nudity is a value, or at least not a cost, American technology can actually provide a gas-fired sauna bath at an energy cost below that of the Japanese or Finnish system. Including a short, hot shower in addition to the gas-fired sauna, the energy cost per person becomes one-sixth that of the big electrically-heated bath, and about one-third that of a big gas-heated tub bath. The flexible cultural attitudes common in communes, combined with good technology, offer the commune considerable energy savings beyond those observed in the presence of conventional technology.

CONCLUSIONS ON THE NEW LIFE-STYLE

We have seen that attitudes toward energy use in the home can be tightly coupled to cultural attitudes about the nature of comfort and leisure as well as cultural perceptions of the desirability of hand labor. For example, it is a common insight that when dishwashing is shared by both sexes, the labor is lighter because it is not a symbol of oppression. Thus, in many homes, a cultural change in attitudes toward sex roles may end up having the environmental effect of making hand-

Table 4 Energy Use and Bath Technology

Bath Style	Hot Water (gallons)	Resource (BTU)	Persons Served	BTU per Person	Comment
Big tub bath (electric water heater)	30	81,000	1	81,000	Electric power generation and transmission efficiency of 27.3%
Big tub bath (gas water heater)	30	35,000	1	35,000	Assumes tank temp. increase of 87°F. and 62% efficient gas water heater
Small tub bath	15	17,500	1	17,500	Gas water heater
Five-minute light shower	5	5,850	1	5,850	Gas water heater
Wood-fired Japanese bath and one tub laundry	46	86,300	5	17,300	Doing tub wash after bath saves 27,700 BTU
American machine load laundry	24	27,700	1	27,700	Gas water heater
Electric sauna and cold showers		82,000 (space heat)	5	16,400	
Electric sauna and one hot shower per person	25	29,300 (water) 82,000 (space) 111,300	5	22,200	300 cu. ft. room with 7.5 kilowatt heater operating 0.88 hours out of 1.25 hours (MacLevy Products Corp., Elmhurst, N.Y.)

Gasfired sauna and cold showers	—	22,000	5	4,400	350 cu. ft.; 25,000 BTU/hr. unit operating 0.88 hours out of 1.25 hours (Viking Sauna, San Jose, Ca.)
Gas sauna and one hot shower per person	25	29,300 (water) 22,000 (space) 61,300	5	12,300	
Solar heated bath	30	—	1	—	Capital investment of from $500 to $800 for solar collector and elevated insulated storage tank

dishwashing feasible, reducing energy used for hot water for dishwashing by a factor of four and eliminating the use of strong machine dishwashing detergents.

Many of today's young are interested in eastern religions, and such teachings may have a profound influence on how individuals view labor. For instance, Zen teaches that the insights possible through disciplined meditation are also possible when the disciplined alertness of meditation is carried to manual labor, and ultimately to every corner of one's life, with the preferred states of mind being possible through manual labor. Thus, a chore like dishwashing loses the onus which some sectors of our culture attach to manual as opposed to mental or spiritual endeavors. Of course, this insight is also available through such native sources as Henry David Thoreau and some American Indian patriarchs.

From this viewpoint the claim that household labor saving devices "liberate man" may be viewed with skepticism by followers of the new life-styles. The dishwasher and the vacuum cleaner (for the wall to wall carpet) may ultimately be discarded by some as encumbrances. This concept may be more applicable to a commune with six adults and two children than to a nuclear household with two adults and six children.

In considering personal energy consumption and life-style, it is convenient to examine the interplay between social structure, cultural expectation, and technology. The material goods enjoyed by the Minneapolis commune sample were similar to those enjoyed by most middle-class households, with the exception that dishwashers and air-conditioners were entirely absent from the twelve communes visited. The low personal energy budgets appeared to be basically economies of scale. However, the same cultural flexibility that brought the 131 commune members into voluntary associations might be an indication that individual adherents of the new life-style would consider other adaptations, if they were available.

About 90% of the average American domestic energy budget is devoted to cooking, space heating, and water heating. Considerable savings in energy in those areas are possible through the use of localized heating combined with cooking, and baths designed for community use, and more clothing. In addition, adaptations to greater variations in home temperatures seem promising, though poorly researched at this time.

Thus, in an era when Americans are staggering before the environmental and economic problems associated with a rapidly expanding industrial economy, this paper is intended to suggest that the assumptions commonly made about domestic energy needs are not based on absolutes. They are a function of decisions of taste and culture, as well as the absolutes of physiology and technology.

A SHORT SOCIAL PERSPECTIVE ON THE URBAN LIFE-STYLE

It is clear from these exercises that many of the alterations in life-style which might yield home energy savings would require the abandoning of popular aspirations for appliances by the disadvantaged. Change in appliance ownership now is more rapid among blacks than among whites, reflecting the fact that the white appliance market is more nearly saturated than that of the blacks (Table 5).

In considering the life-style of the disadvantaged, one must recognize that the dearth of major appliances is accompanied by basic needs for housing and transportation.

The destruction of the ecological and social niche of many rural poor through pesticides, herbicides, and harvesting machines described by John Hatch (1971) and Michael Perelman (Chapter 4), has also resulted in mass migrations to the cities, where the poor have the indignation of living in the swill committed by factory stacks feeding junk products into a mass flow economy which cannot furnish satisfactory levels of employment (Hare, 1970).

Table 5 Percent of Households Owning Selected Durables January 1967 and July 1971

	Black			White		
	'67	'71	% Change	'67	'71	% Change
Automobiles:						
One or more	51	52.5	3	83	83.1	0.1
Two or more	10	13.2	32	28	31.5	12.4
Three or more		0.5			5.2	
One or more recent model automobiles (1966–1967)	6			17		
Durables:						
Black and white TV	83	89.6	8	87	96.0	10
Color TV	5	23.6	470	16	45.0	28
Dishwasher	3	3.7	23	13	20.5	58
Room Air Conditioner	7	11.1	59	19	23.3	22

Source: U.S. Department of Commerce, Bureau of the Census. Technical Studies, Series P-23, No. 1-25, Ref. C 3.186: (for 1967 Data); and Survey of Consumer Buying Expectations, July 1971 (unpublished data).

In 1970, 10 million American housing units were either crowded, sub-standard, or both (RHA, April, 1972). Dr. A. Allan Bates, Director, Office of Standards Policy, U.S. Department of Commerce, argues that the poor who usually occupy the worst housing cannot afford to live in the kind of housing our economic system produces unless it is subsidized and that subsidization simply does not produce the kind of low cost housing that could eliminate the housing shortage in this country.

Pointing to the experiences of Western and Eastern Europe, he suggests that industrialized housing techniques should be used to create massive microdistricts which would be relatively self-contained in terms of work, school, shopping, and services, thus easing the life of the low income wage earner, and providing a low impact style of life (Bates, 1969). However, the possibility of aluminum cities, buildings of stacked trailer houses, and disposable houses mentioned in the research synopsis by Bates suggests that we may be on the brink of massive programs which will change the quality and style of human life, and the human environment. In this context, we may gain some insights from the history of urbanization. The densely populated old central cities are one of the few strongholds of the life-style dependent on light impact mass transportation. Also, mean family auto mileage in old central cities was 3900 miles per auto per year, compared to a national average of 12,900 miles per household vehicle, a pattern which is not due to differences in income (Lansing and Hendricks, 1967). A similar inverse relationship to population concentration exists for washing machine ownership (Bureau of Census Consumer Reports, 1971). Such empirical evidence suggests that higher residential densities, *if they could be made livable in our culture,* might yield resource advantages, especially through community facilities, community cultural activities, etc.

Curiously, hope for new urban communities expressed by construction men such as A. Allan Bates is shared by some experts from the energy industries. Sam E. Beall and Arthur J. Miller of ORNL have suggested small-scale high efficiency "total energy systems" for housing developments in the size range suggested by Bates:

> Since new communities tend to grow in modules of several hundred or a few thousand dwelling units, we are now evaluating for HUD the concept of energy system modules. Prime movers such as gas turbines gas engines or diesel engines, etc., would be used to supply energy in the form of electricity, heat, and air conditioning. The heat intensive portions of the system would make use of exhaust and coolant heat from generation of electricity by the prime mover. These local energy systems or "total energy" systems are currently used in some industrial plants, commercial buildings, and apartment houses. They have a low

electrical generating efficiency, but when there is an adequate recovery and demand for heat, they have a high efficiency of energy utilization. (Beall and Miller, 1971)

Thus, in addition to facilitating economies in construction through industrial building techniques, and economies in transportation through community design, the urban life-style apparently offers the possibility of substantial reductions in residential energy consumption for heating and cooling. However, such developments could be futile if not accompanied by a rationalization of energy use in large buildings along the lines documented by Stein in Chapter 3.

The full interweave of life-style urban non-planning, and energy consumption has only been hinted at here. Since many advocates of new life-styles have failed to demonstrate how personal preferences could affect the contrary movement of our economy, their work has been dismissed as the "cultural demystification" of "commodity fetishism" (Gintis, 1972). However, the possibility remains that the new life-style phenomena, close as it is to market issues, could be resocialization preceeding institutional change (Mitchell, 1971).

This chapter has not addressed the sociological and economic considerations behind the prevalent life-style in the United States. Thus, it is primarily a suggestive and technical effort showing that "The New Life-Style," lean and communal, makes only a fraction of the per capita energy demand on the economy than that which is now prevalent. In addition, the material on heating and transportation suggests that economies are realizable through the use of low energy technologies and urban advantages of scale. The authors hope the public will give due consideration to these adaptations as they plan for the culture of the 21st century.

REFERENCES

American Society of Heating, Refrigeration and Air Conditioning Engineers, 1972. *Handbook of fundamentals.* New York, p. 136.

Automobile Manufacturers Association, 1971. *1971 auto facts and figures.* New York, p. 22.

Banerji, S. K., 1959. Comfort zone. *Climate, environment and health.* National Institute of Sciences of India, New Dehli, p. 28.

Basu, N. M., 1959. Life in cold and hot climates. *Climate, environment and health.* National Institute of Sciences of India, New Dehli, pp. 32–35.

Bates, A. Allen, 1969. Testimony before the Subcommittee on Urban Affairs of the Joint Economic Committee of the Congress of the U.S., Hearings on Industrialized Housing, Apr., 1969.

Beall, Sam, Jr., and Arthur J. Miller, 1971. The use of heat as well as electricity from electricity generating systems. A paper given at the AAAS symposium on the Energy Crisis, Philadelphia, Dec.

Brian, Margaret. Northern State Power, personal communication, 1972.

Brooks, F. A., 1955. Use of solar energy for heating water. In Daniels, Farrington, and Jabaffie, eds. *Solar energy research*, Madison, Wisconsin.

Bureau of the Census, 1971. 1970 Census of Housing, Minneapolis-St. Paul Urbanized Area, Block Statistics, Department of Commerce, Aug., Table 1.

Bureau of Statistics, Office of the Prime Minister of Japan, 1968. *Japan statistical yearbook, 1968*. Tokyo.

———, 1971. *Japan statistical yearbook, 1970*. Tokyo.

Commoner, Barry, and Michael Corr, 1971. Power Consumption and Human Welfare in Industry, Commerce and the Home, paper for the Energy Crisis Symposium, AAAS Annual Convention, Philadelphia, December, 1971. Survey of Consumer Buying Expectations, U.S. Bureau of Census, unpublished data, July, 1971.

Commoner, Barry, et al., 1971. The causes of pollution. *Environment* Apr.

Corr, Michael, and Dan MacLeod, 1972. Home energy consumption as a function of life style. Electric Power Consumption and Human Welfare, AAAS/CEA Power Study Group, Aug. 11, review edition.

Evans-Wentz. W. Y., 1970. *Tibetan yoga and secret doctrines*. New York: Oxford Univ. Press, p. 160.

Fanger, P. O., 1970. *Thermal comfort*. Copenhagen: Danish Technical Press, p. 49.

Fitch, J. M., 1960. Primitive architecture and climate. *Sci. Am.*, Dec.

Garfield, Viola E., et al., 1961. The wolf and the raven. Seattle: Univ. of Washington Press, p. 77.

Gintis, Herbert, 1972. Activism and counter culture: The dialectics of consciousness in the corporate state. *Telos*, no. 12 (Summer).

Goodman, Paul, and Percival Goodman. *Communitas: Means of livelihood and ways of life*. New York: Random House.

Hare, Nathan, 1970. Black ecology. *The Black Scholar* Apr.

Hatch, John, 1971. From crisis to disaster: An account of the struggles of black farm laborers in the United States. Paper presented at AAAS Conference on Environmental Sciences and International Development.

Hawkins, Harold M., 1947. *Domestic solar water heating in Florida*. Gainesville, Florida Engineering and Industrial Experiment Station; College of Engineering, Univ. of Florida, Bulletin No. 18, Sept.

Hirst, Eric, 1972. *Energy consumption for transportation in the U.S.*

———, 1972. *Electric utility advertising and the environment*. ORNL, Apr.

Kraus, William, 1972. Personal communication, May.

Lansing, John B., and Gary Hendricks, 1967. *Automobile ownership and residential density*. Ann Arbor, Institute for Social Research, Univ. of Michigan, p. 24.

Luszczynski, K. H., 1972. Lost power. *Environment* Apr.

Mitchell, Juliet, 1971. *Woman's estate*. New York: Pantheon, p. 32.

National Industrial Pollution Control Council, 1971. *The disposal of major appliances,* June.

New, R. E., 1966. The electric utility looks at the domestic appliance and electric heat load. IEEE Industry and General Applications Group Annual Meeting, Chicago, Illinois, Oct. 3–6.

Office of Science and Technology. *Patterns of energy consumption in the United States* (unpublished). Excerpts in Executive Office of the President, Office of Emergency Preparedness, *The potential for energy conservation.* U.S. Government Printing Office, SN-4102-00009, Oct., 1972.

Rural Housing Alliance. *Low income housing bulletin* Apr., 1972.

Scientific American, 1970. *Input output table of the United States economy.* New York.

Seattle City Light. *Operating costs of your electric servants.* 3-67-50M.

Senate Public Works Committee, 1970. Hearings on disposal of junked and abandoned vehicles. Y4, p. 96/10:J96 (1970).

Shakow, Don. Ph.D. diss., Economics, Berkeley. Personal communication, 1972.

Snyder, Gary, 1969. Why tribe, and Buddhism and the coming revolution. In *Earth household.* New York: New Directions.

———, 1972. *New York Times* Jan. 12.

Stelzer, Irwin M., 1971. Zero-growth power advocates ignore effect on the poor. *Electric World* 175, No. 6: 147, 148.

Trane, R. N., 1923. Heating practices in England. *Domestic Sci.* Feb.: 364.

The Whole Earth Catalog, 1970. Portola Institute, Inc.

Walker, Connecticut, 1972. Hitchhiking in Poland—Licensed, safe, profitable. *Parade Magazine, St. Louis Post-Dispatch,* July 30, 1972.

HERMAN E. DALY

Electric Power, Employment, and Economic Growth

The salient fact about electric power production is that it is growing at 7% annually, doubling every ten years. Even ardent supporters of energy growth recognize the absurdity of projecting such a rate very far into the future, lest the entire environment be transformed into electric power and generators thereof. For example, Mr. John W. Simpson, president of Westinghouse Power Systems Company says the following:

> At some point in time, well before power plants crowd us into the ocean or we change the climate of the earth through the rejection of heat into the atmosphere, we will reach an energy plateau, a level at which our technology will have placed us through the super-efficient utilization of power. In other words we will have learned to do so much more with so much less power that more power plants as we know them today probably would be superfluous. (*Congressional Record*, March 8, 1971, p. E1566).

How far in the future is this "energy plateau"? Mr. Simpson says only that, "Well before that hypothetical 200 years from now, we will have seen a leveling off in the rate of increase of power consumption." Within 200 years he foresees a "leveling off in the rate of increase." Presumably this means a zero rate of increase rather than some constant positive rate. But the important concept of an energy plateau is left vague, and for the forseeable future (30–50 years?) he argues for more power plants as we now know them. Congressman Holifield (Joint Committee on Atomic Energy) puts a minimum figure of 30 years on the "forseeable future" in stating that ". . . the doubling factor every decade for the next 30 years is basic to our standard of living, now and in the future" (Holifield, 1971). According to this point of view, one can make the loose inference that, at the minimum, eight times our present electric power output might be enough—then we can begin to talk seriously about an energy plateau, not before. But certainly we could not abruptly decelerate from 7% to zero growth immediately at the end of 30 years. It would probably be necessary to double at least once more over a longer period, in order to overcome the inertia of growth without shaking up the economy too badly. Thus the "energy plateau" appears to be at least sixteen times the present output! Furthermore we are to grow as rapidly as we can (at least at 7%) for the next 30 years, which is to say the next generation, or sufficiently far in the future not to affect the capitalized market values of electric utility "growth" stocks. After us, the energy plateau—or perhaps the deluge?

For the regulated electric utilities, growth and profitability are closely related. The regulated "fair rate of return on capital" has been interpreted by the courts to involve three criteria. The rate of return must: (1) be commensurate with returns on investments in other enterprises having commensurate risks, (2) be sufficient to insure the financial integrity of the enterprise, (3) be able to maintain its credit and attract capital. The relevant criterion is the last. In order to attract capital to finance the rapid expansion of capacity the rate of return must be *higher* than in other investments of equal risk. The need for rapid growth, if established to the satisfaction of regulatory agencies, must lead to a higher allowable rate of return. Higher rates have already been granted in some cases and many petitions are pending. Furthermore, even if the regulated rate of return is held constant the utilities have abundant incentive to expand productive capacity in order to increase the base to which the regulated rate is applied, and thus increase total profits. If the rate of return must be raised to attract capital, so much the better. But in order to expand and to raise the rate, the regulatory commission must be convinced of the need for expansion. To convince the commissions the utilities must be able to point to a "shortage." To get a shortage demand must increase. To increase demand it is necessary to advertise and give lower rates for large quantities. Perhaps this explains the curious fact that the electric utilities, a public monopoly, spend eight dollars on advertising for every one dollar spent on R&D (Fabricant and Hallman, 1971). Another incentive to growth beyond the optimum is that a large part of the cost of increased electric power is the social cost of environmental deterioration, which is paid by society in general, rather than by the parties directly responsible for the costs. Although one is not accustomed to thinking of electric utilities as a "growth stock," Standard and Poor's Corp. now informs the investor that "electric power is not only the backbone of the American economy(!), it is also a vigorous growth industry." (Standard and Poor's Corp., 1971).

Electric power is now used to burn neon signs in the daytime, to run electric toothbrushes and moveable sidewalks. With so many trivial uses of electricity in current vogue, one wonders just what we will do with twice as much in 1980, four times as much in 1990, etc. It is not my intention to belittle the tremendous importance of electrical energy or of production in general. The French economist de Jouvenel (1971) has put the issue very well:

> But if I do not at all object to the much enhanced status of Production, I may point out that Production has come to embrace so much that it would be foolish to grant any and every productive activity the moral benefit of an earnestness not to be found in so-called "non-productive activities." When popular newspapers propose to bring out their comic strips in color, I find it hard to regard such "progress in production" as something more earnest than planting flowering shrubs along the highways. I am quite willing to regard poetry as a frivolous occupation as against the tilling of the soil but not as against the composing of advertisements. When organizers of production have to relieve a situation of hunger, efficiency is the one and only virtue. But when this virtue has been thoroughly developed and comes to be applied to less and less vital objects, the question surely arises of the right choice of objects.

This declining intensity of the needs which increased production could satisfy is recognized as important by advocates of growth also. They regard it, however, as an *obstacle to be overcome* in order to increase the flow of output and employment. For example, economist Joseph N. Froomkin believes:

> The big challenge to the U.S. and other Western economies is to bring out cheaply, through automation or otherwise, new products that will tempt the consumer's jaded tastes. These new products in turn will stimulate investment.

Again, if the "big challenge" presently is to tempt the consumer's "jaded tastes" what are we going to do with sixteen times as much electricity? Population growth will absorb only a small and declining portion. The big increase will be, and has recently been, in per capita consumption. Perhaps we must all learn to imitate President Nixon's idiosyncracy of turning the air-conditioner on full blast so that he can enjoy warming himself by the fireplace in mid-summer![1] This is rather logical. Central heating in the winter makes it too warm inside to enjoy the hearth fire, which man has loved from earliest times. In the summer air-conditioning makes it too cold inside, and thus possible to

[1]Interview with Tricia Nixon, reported on national news.

enjoy a fire. Such self-cancelling uses are capable of absorbing great amounts of electric power, and perhaps of jading consumer's tastes to the very limit. But is such "production" more earnest than weeding a flower garden? The question of electric power for the poor is considered later.

The above considerations raise doubts about whether the impossibility of growing forever at 7% is really taken seriously, or if the notion of an "energy plateau" is not just mentioned in passing in order to fend off any easy *reductio ad absurdam* by opponents. At least the concept is left undeveloped. These doubts are increased as one reads the four main arguments by which the growth position has been defended in the media and in congressional hearings. These arguments are stated below, critically examined, and found to be grossly deficient both in their logic and premises. The growth advocates state that we must have continued rapid growth in energy in order to:

A) Increase the standard of living (level of consumption) of the present population, and extend that rising standard of living to some 70–100 million more Americans by the year 2000.
B) Clean up the pollution and recycle the wastes which inevitably result from current and increased levels of production and consumption, and which have accumulated from past producton and consumption.
C) Maintain an acceptable level of employment, since (1) labor is used directly in the production of electric power, and more importantly (2) power indirectly affects employment since it is an input in most production processes which also use labor, and it is a necessary complement to many consumer goods whose production employs people.
D) Produce the devices which give us the military defense and deterrent capacity "upon which world peace depends."

Our attention will mainly be directed to reasons A and C, but it is important by way of introduction to realize that the four reasons are part of a whole point of view and are not independent of one another. The tacit, unifying point of view, the unarticulated preconception, is that of open-ended growth. Although open-ended growth is sometimes denied by refusing to project the growth of the energy sector beyond one generation, it remains an implicit assumption of each of the four arguments for the "temporary" necessity of continued rapid growth. Reason A postulates the goal of an increasing level of consumption with no mention of a "consumption plateau." But even the growth-oriented President's Council of Economic Advisors has told us that "growth of GNP has its costs, and beyond some point they are not worth paying" (*Economic Report of the President,* 1971). Such a point does not enter into reason A. Although there is a growing body of opinion that suspects that in the United States we may have already passed the critical point, reason A fails to recognize even the existence of such a point. As total output continues to grow and grow so does total waste and pollution. Reason B, more energy needed to clean up, is thus guaranteed by reason A, as well as by a backlog of accumulated pollution. Since matter and energy can be neither created nor destroyed, the more we "consume," the more waste we must clean up. Thus the first round of matter-energy degradation (production) requires a second round to clean up after the first, and in turn the second round of matter-energy degradation requires a third round to clean up after it, etc. There is an infinite series of clean-up uses of energy, although the terms of the series, like the dirt remaining after each sweep into the dust pan, probably diminish rapidly. Nevertheless there is something of a vicious circle in burning coal to power industry, and then having to burn more coal to produce the energy to blow away the smog, to clean up the buildings, to run the washing machines and dryers more hours to clean white shirts more often, to produce more washers, dryers, and white shirts to replace the more rapid wearing out of these items resulting from greater use, etc. This degradation aspect of production is of vital importance because what is ultimately scarce in the physical world is not energy but low entropy. The more energy we use, even for the purpose of cleaning up, the more rapidly we use up the stock of low entropy—i.e., the more rapidly we degrade the total environment. Thus reason B derives from reason A and also reinforces reason A by falsely suggesting that cleaning up *cancels* the environmental cost of production growth. Cleaning up is itself a *part* of the environmental cost of production, and while necessary, is certainly not an activity to be pursued beyond some limits. At some point the degradation cost of energy needed to clean up will be greater than the cost of the pollution which is to be cleaned up. Reason C argues for more energy in order to provide more employment. But inanimate energy is a *substitute* for human labor, as well as a complement. With a given total output an increasing supply of inanimate power will surely be used to substitute for labor (what else could it be used for?) and will result in higher productivity per worker, but *less* employment. The reduced hours of employment would probably result in fewer men employed, although it could result in the same number of men working fewer hours. But since no mention is made of reducing the hours worked per man, and since this can increase men employed independently of changes in energy use, it is clear that argument C assumes that total output is growing at a rate faster than the rate at which power substitutes for labor. In other words we are back to reason A and unlimited growth in total product. Reason D appeals to national "defense" or deterrence needs for energy. But deterrence is not a thing or even a state. It is a self-escalating process of open-ended growth, a balance of power equation in which both sides continuously grow in tandem until the arms of the balance break. It meshes perfectly with the "growthmania" assumption of reason A. The more the military depletes the civilian economy of resources and technical personnel, the lower will be growth in productivity of the civilian economy. The slower the growth in productivity the greater must be the growth in raw material and energy inputs to maintain a given rate of increase in product. The technical depletion of the civilian economy by burgeoning military "needs" has reduced possibilities of the one kind of growth which we all welcome, growth in productivity, or less input per output. It makes the "energy plateau" much harder to attain.

For the above reasons one cannot escape the strong impression that the fleeting mention of limits or plateaus is nothing but a base-touching digression, a debating parry, and that the paradigm underlying the four arguments remains one of "growthmania." In addition to the absurdities of projecting high growth rates of electric power into the future, we must add the absurdity of projecting high growth rates of total output, if we are to take the four arguments at their logical face value. To paraphrase Mr. Simpson, it seems that at some time in the future, well before power plants crowd us into the ocean, we will have already been crowded into an already polluted ocean by our commodities, their effluents, and their corpses! Our economy, a living thing, is succeeding too well in the imperialist drive noted by Bertrand Russell to "transform as much as possible of the environment into itself"—i.e., into commodities and people. The electric power industry has simply been more successful in doing what all other industries are striving to do—namely to grow, and when possible to avoid paying the full social cost of the increased production. This points to serious irrationalities at the overall systems level of our economy, some radical implications of which will be treated in Section II.

Further attention must be given (in Section II) to reason C, the employment argument, since it is both particularly influential in a period of unemployment, and particularly fallacious. The key argument, however, is A, open-ended growth in consumption. If A falls the other arguments fall with it, so it will merit our special attention in Section II.

I. ENERGY AND EMPLOYMENT

Bernard de Mandeville in the "Fable of the Bees" argued in favor of the most extravagant luxury consumption by the upper class on the grounds that it provided employment for the poor. Frederick Bastiat in his "Petition of the Candlemakers—" argued in favor of blocking out sunlight from all houses and buildings in order to increase employment among candlemakers, who in turn, through their increased expenditures, would stimulate employment in whaling, matchmaking, wick production, and in effect all industries, to the immense benefit of all the classes of the realm. These two gentlemen were of course speaking tongue-in-cheek with the object of satirizing some of the contemporaries who were making analogous arguments in all special-pleading sincerity. Whether the employment arguments of today's electric power interests belong to the same category I will leave for the reader to judge, presuming only to list a number of considerations which might be relevant in making such a judgment.

(1) Physical labor is at the margin a negative factor in the enjoyment of life. Leisure is at the margin a positive factor in the enjoyment of life. Thousands of years ago man domesticated animals to relieve his toil by substituting non-human for human animate energy in production. Later man "domesticated" inanimate energy and again substituted it for human energy and leisure previously sacrificed to production. It would appear that historically the dominant relation between human and non-human energy is that of substitution. The more of one, the less of the other, a negative correlation. Yet we find in fact a positive correlation. Historically the number of men employed has increased as non-human energy use has increased. Therefore one is tempted to argue that more employment necessitates more use of inanimate energy. But this crude empiricism will not stand up to simple rational analysis. The observed positive correlation between non-human energy and employment means only that energy use and employment are both correlated to some third factors which have been increasing historically, namely total output and total population. The positive indirect correlations via the growing third factors outweighed the negative direct correlation between energy and employment themselves. For a constant level of output and population we know that the correlation between energy and employment is bound to be negative for as we previously asked, what conceivable use of non-human energy is there besides either substituting for human labor in producing the same or a smaller quantity of goods, or complementing human labor in order to produce a greater quantity of goods? If we rule out the second we are left with the first. If we insist on more employment and more non-human energy we must also insist on more output. In the real world both substitute and complementary uses of inanimate energy are occurring simultaneously. Only if total product growth creates more jobs than are eli-

minated by the growth of non-human energy will employment increase. This is self-evident.

Congressman Chet Holifield, a supporter of energy growth, nearly proved too much in extolling the importance of energy.

"Our standard of living is the highest of any nation in the world because we use the highest ratio of mechanical power versus manpower."

"One man can do the work of 350 men." (Holifield, 1971).

This is a clear recognition of the overwhelming importance of increasing substitution of non-human for human energy as the major factor in increasing the standard of living. Furthermore, for one man to do the work of 350 men means that 349 men must do something else. If they are employed doing something else total product must increase. If total product does not increase they must be unemployed. Clearly what is responsible for increasing total employment is the increase in total product, not the increase in inanimate power consumption which in fact, taken by itself, must *decrease* employment.

(2) As non-human energy has replaced human energy in agriculture it has not been possible to provide much employment in this sector, even with vast increases in total agricultural product. Employment was provided by growth in the new sector of industry. As non-human energy replaces men in industry, even with large increases in total output of the industrial sector, the major source of new employment has been the "new" service sector rather than industry. The service sector's share of total employment has grown from approximately 40% in 1929 to over 55% in 1967. Of the total net increment of 14 million jobs between 1947 and 1965, the service sector accounted for an increase of 13 million, industry an increase of 4 million, and agriculture a decrease of 3 million. (See Fuchs, 1968.) Services, the least energy-using of the three sectors, have provided nearly the entire net increase in employment since 1947. Service institutions such as banks, hospitals, retail stores, schools, insurance companies, etc., are the providers of new employment, and their energy requirements are for lighting, space heating and cooling and office machines—perhaps no more energy intensive than the average household, on a per capita basis. The most energy-intensive sectors are also the most highly capitalized and automated, and consequently are not significant providers of new employment. Thus the alleged need for large increases in energy input to provide new jobs seems to presuppose that the average new worker will work in a steel or aluminum plant. This is counter-factual. He will work in a service institution.

What is the reason for the drastic shift of employment toward the service sector? The authoritative study by Victor Fuchs (1968) considers three hypotheses: (1) a more rapid growth of final demand for services, (2) a relative increase in the intermediate demand for services, and (3) a relatively slow increase in output per man in services. Reason (1) could not be very important since the service sector's share of total output was the same in 1965 as it was in 1929, when measured in constant dollars. Measured in current dollars it increased only from 47% to 50%. Hypothesis (2) is examined statistically and found to account for less than 10% of the total change. The major explanation is hypothesis (3), that output per man grew much more slowly in the service sector than in the other sectors. This means that the amount of labor required for a given output fell more rapidly in agriculture and industry than in the service sector. In other words, services employ relatively more people because agriculture and industry need relatively less labor year after year.

But *why* has productivity per man risen more slowly in the services than in industry? Are services inherently less subject to technological improvement than manufacturing? Perhaps so, as Fuchs seems to suggest, but there is another reason which should not be overlooked. The industrial sector of the economy is characterized by much greater monopoly power than is the case with services. This power tends to be "countervailing" with oligopolistic managements offset by monopolistic labor unions. More than half the persons employed in industry are union members, while the service sector is only about 10% unionized (Fuchs, 1968). Unionized labor succeeds in temporarily pushing wages above the marginal revenue product of labor (MRP_L). Management reacts by seeking to reestablish the equality in order to keep short run profits at a maximum.[2] As wages rise the firm would produce less in the short run, which would tend both to increase the product price and the marginal physical product of labor, and consequently the productivity per man measured in value terms. In the long run management reacts to union power by seeking to substitute capital for labor, thus further increasing the productivity of labor and reducing the number of laborers needed. In the service sector, by contrast, we witness neither much unionization nor much monopoly power, since entry into services is not blocked by high initial capital outlays as in industry. Such "cartelization" as does exist in services (physicians, lawyers) in-

[2]As long as the wage is less than the marginal revenue product of labor the firm can increase profit by hiring more laborers. If the wage is greater than the marginal revenue product of labor, profits are increased by hiring fewer laborers (i.e., the last laborers hired cost more than they were worth to the firm). When wage equals the marginal revenue product of labor it is impossible to increase profit by hiring either more or less labor—in other words profit is at a maximum.

volves no conflict between labor and management with the resultant upward pressure on productivity per man. It seems, therefore, that at least a part of the difference in productivity and employment between industry and services has its origin in differential market power rather than in differential susceptibility to technological progress.

One-third of service sector employment is in non-profit institutions where the employment-limiting rule of wage equal to MRP_L loses its rationale from the very beginning. Also the service sector has a much higher incidence of self-employment where the wage equal to MRP_L rule applies with less force and where a decline in employment often takes the work-sharing form of fewer hours worked or less intensive work by the same number of people. Many small entrepreneurs in services are also, however unintentionally, "non-profit institutions" and are providing temporary employment at more than MRP_L while in the process of slowly going out of business. Retail gasoline stations are a case in point. Finally, for many services output is so hard to measure that it is estimated for national income purposes by the value of input (e.g., government services), with the result that productivity loses all meaning. Often productivity in services, even when it can be roughly measured, depends as much on the consumer as on the producer. For example, a physician's true productivity (not the dollar value productivity which equals the fee paid) depends on the patient's ability to supply an accurate medical history, and the teacher's productivity varies according to the interest and intelligence of the student. In sum, the whole notion of productivity in the service sector is often very vague (may depend on consumer as well as producer), or completely meaningless (as when we estimate the value of output by the value of input), or of little consequence in determining employment (as in non-profit institutions, both intentional and unintentional). The upshot is that service sector employment is much less strictly limited by productivity considerations than is industrial employment.

On the demand side ever-rising standards of minimum consumption levels for certain services (education, medical care, legal counsel, insurance, government welfare services) have helped to keep demand up with supply and to make possible growing employment in these areas. But there is a limit to the amount of services people will voluntarily consume. Much service consumption is already involuntary—e.g., compulsory schooling, quasi-compulsory college attendance in order to get a job, compulsory insurance and retirement services and related medical exams, compulsory psychiatric care for the "mentally ill," etc. How far such forced consumption is justifiable is now becoming a hotly debated issue. Books are written on "Deschooling Society" and on "How to Avoid Probate" and the associated legal fees. The notion of "legal insanity" and forced consumption of psychiatric services is also under attack.

It has long been argued that in underdeveloped countries the service sector, particularly government services, serves distribution more than it serves production, by acting as a sponge to absorb the unemployed. Might not the same apply to the United States?

(3) One must recognize of course that men are employed directly in energy extraction, refining, pipeline transporation, and utilities. This employment amounted to 2.6% of total employment in 1947, 2.3% in 1955, and 1.8% in 1965. The energy sector accounted for a much larger percentage of total investment; 16.1% in 1947, 22.0% in 1955, and 19.4% in 1965. The percentage of GNP originating in the energy sector was 4.5% in 1947, 4.7% in 1955, and 3.8% in 1965. Note that the share of energy in investment is very high relative to its share in total product. Note also that energy's share in employment is low relative to its share in total product. Thus it requires a relatively large investment in energy to produce a relatively small amount of employment. Energy transformation is probably the most capital-intensive, labor-saving sector of the economy. Direct employment from its growth will be very small.

(4) Indirect employment effects are another matter, and have been touched on already. We may consider two kinds of indirect employment effects: those arising from interdependence on the supply side, and those arising from the demand side. On the supply side it is undeniable that energy is a necessary input to nearly all sectors. But it does not follow that increasing the production of energy will increase the number of jobs in these sectors. To the extent that increased supply lowers the price of energy the result is more likely to be a reduction in employment as cheaper energy is substituted for human labor. Nor does it follow that other sectors cannot expand employment without an increase in energy input, except in the special case of fixed coefficients of production. In general, therefore, an increase of enrgey input to the supply side of the economy is neither a sufficient nor a necessary condition for increasing employment. Furthermore the same supply interdependence arguments would apply equally to many other sectors, not just energy. Minerals are necessary inputs to most sectors, so the jobs of all people in all sectors depend on minerals. Likewise for water. For some production processes a 3/4" no. 8 counter-sunk chromium plated brass wood screw is a necessary input. Therefore the jobs of all people in those sectors depend on this type of screw. And so on, all of which proves nothing.

On the demand side there are indirect employment effects resulting from new investment in the energy sector. As we have seen the energy sector does account for a significant amount of new investment, some 20%. This high percentage of investment results from rapid growth and high capital intensity. These investment expenditures generate a chain of respending or "multiplier effects" throughout the economy, stimulating output and employment (and inflation). But once again this Keynesian argument is perfectly general and can be applied to new investment in any sector. Indeed, it can be applied to increased expenditure of any kind (increased welfare payments, unemployment compensation, or military spending), not just investment. Arguments appealing to indirect employment effects, either via supply or demand, are therefore obvious cases of special pleading—they apply to all sectors, not just energy. Such arguments, when they rely on investment as they usually do, are also premised on open-ended growth in total product (reason A).

(5) An occasional variant of the employment argument is what might be called the "power for the poor" argument. To quote Mr. Simpson (1971) again.

> If we were to freeze our rate of power generation and consumption, we would effectively deny those millions who are below the average any chance to improve their lot in the future.

This statement is a non-sequitur. The lot of those below the average does not depend on electric power alone, and even if it did their lot could still be improved by cutting down the consumption of those who are above average—evidently a thought too horrible to contemplate! No evidence is presented to the effect that the increment in power would go largely to the poor. There is reason to expect just the opposite.

Consider the same argument as expressed in the "Energy Marketing" section of *Electrical World* (March 15, 1971). In an article entitled "Zero Growth Power Advocates Ignore Effect on the Poor" it is argued that zero growth in power is a policy "which, by a sort of grandfather clause, would seek to prevent those of lower income and darker skin from sharing in the not-inconsiderable comforts and conveniences afforded by modern electric appliances . . ." As "evidence" we are presented with the interesting data in Table 1 which shows that a much larger percentage of the wealthy and white own major electric appliances than is the case for the low income and non-white. The "conclusion" that the reader is encouraged to draw from these statistics is that, "future growth will consist of gradual accumulation by lower income families of the more modern appliances first accumulated by their wealthier cousins." But if the latter comes about it will be because of general economic growth and the resulting growth in employment, not from growth in electric power, which taken by itself is

Table 1 Appliance Ownership by Income and Race (New York City) % Ownership

Income	Air-Conditioner	Dish Washer	Vacuum Cleaner
$ 0–3,000	12%	1%	40%
3–5,000	11%	2%	46%
5–10,000	28%	5%	60%
10–15,000	48%	15%	89%
15,000 +	58%	36%	95%
White	40%	12%	81%
Non-white	8%	3%	41%

Source: Electric World March 15, 1971, p. 148.

likely to have the opposite effect, as we have already seen. More importantly the conclusion which follows from the fact that the rich own relatively more appliances than the poor is that the rich use relatively more electricity than the poor. How can the poor use much more electricity without buying air-conditioners, dish washers, vacuum cleaners, frost-free refrigerators, washers, dryers, color TVs, etc? And how can they buy these appliances if they are poor? If the increment in electric power were truly directed to the poor, one would expect to find the energy marketing people striving to promote mainly simple, low-cost appliances with low operating and maintenance costs, designed to be advantageous to the poor. One would expect to see R&D receive eight times as much as advertising instead of one-eighth as much, as is now the case. One would also expect to see rate discrimination in favor of the small, rather than the large, users. That the energy marketeers prefer to concentrate on production for those who can pay high prices is obvious from the articles appearing in the "Energy Marketing" section of *Electrical World.* For example, consider the other two articles in this section of the same issue. One extols the virtues of a moving sidewalk projected for Boston, the object of which is to convey consumers from the parking area to the shopping area more quickly. It would seem that since conveyor belts have speeded up production, bringing the commodity to the market more rapidly, we must also speed up consumption by putting the consumer on conveyor belts. A moving sidewalk will surely be a boon to the ghettos! The other new dimension in energy marketing is the Gold Medallion home, all electric so as to avoid pollution (!), with central vacuum systems and electronic security systems, the latter no doubt to safeguard the other electric appliances of the poor! Sandwiched between these two articles on the latest technological advance in trivializing energy consumption to "tempt the jaded tastes" of the affluent, the article defending the poor against the selfishness of the zero growth advocates raises suspicions of trumpery. Needless to add, no consideration is given to the fact that power production pollutes, and that while the benefits of power go mainly to the rich, the external costs fall mainly on the poor. For example, in Baton Rouge, where I live, the oil refineries and petrochemical and aluminum industries are located mainly in North Baton Rouge, which is the low income end of town, and which receives the most pollution. The executives of these companies live in southeast Baton Rouge. They drive to work along a north to southeast freeway in an air-conditioned car and work in an air-conditioned office. The manual laborers live in north Baton Rouge, drive with windows open through narrow streets to work in a polluted plant. The average skin color in north Baton Rouge is much darker than in southeast Baton Rouge. The external costs of energy production are distributed regressively—the poor pay more. And this is true not only in Baton Rouge.

(6) It should be clear that growth in efficiency is not being questioned. If we can get more service from the same physical fund, that is all to the good. But increasing the physical fund is clearly a limited process. No one objects to getting "more kicks per kilowatt" or to more kilowatts per BTU of primary fuel. But if kilowatts keep on growing we will witness a decline in the ratio of kicks (satisfaction) per kilowatt. For many years we had a falling trend in the ratio of energy used to GNP. But since 1966 the actual ratio has risen sharply and in 1970 was 13% greater than it would have been had the pre-1966 trend continued. (See Figure 1). Thus a dollar's worth of satisfaction now costs more kilowatts than in 1966, and the new trend seems to be rising. Furthermore, new products and quality improvements notwithstanding, the amount of satisfaction represented by a constant dollar's worth of real GNP is itself probably falling. We satisfy our most pressing wants first, and each extra dollar of income is dedicated to the satisfaction of a less important want. Furthermore real GNP increases currently overstate welfare increases since more and more of what is counted in real GNP is the cost of defending ourselves from the unwanted side effects of production. Therefore the fact that real GNP per capita has been growing means that the incremental dollars represent less satisfaction. With decreases in both satisfaction per dollar and in dollars per kilowatt it is very clear that satisfaction per kilowatt must also have fallen. This does not square very well with Mr. Simpson's prediction that "We will have learned to do so much more with so much less power . . ." It seems instead that we are learning to use more power to do less, at least in terms of the significance of what is done. This by itself does not mean that we have overshot the optimum. Falling marginal benefits of

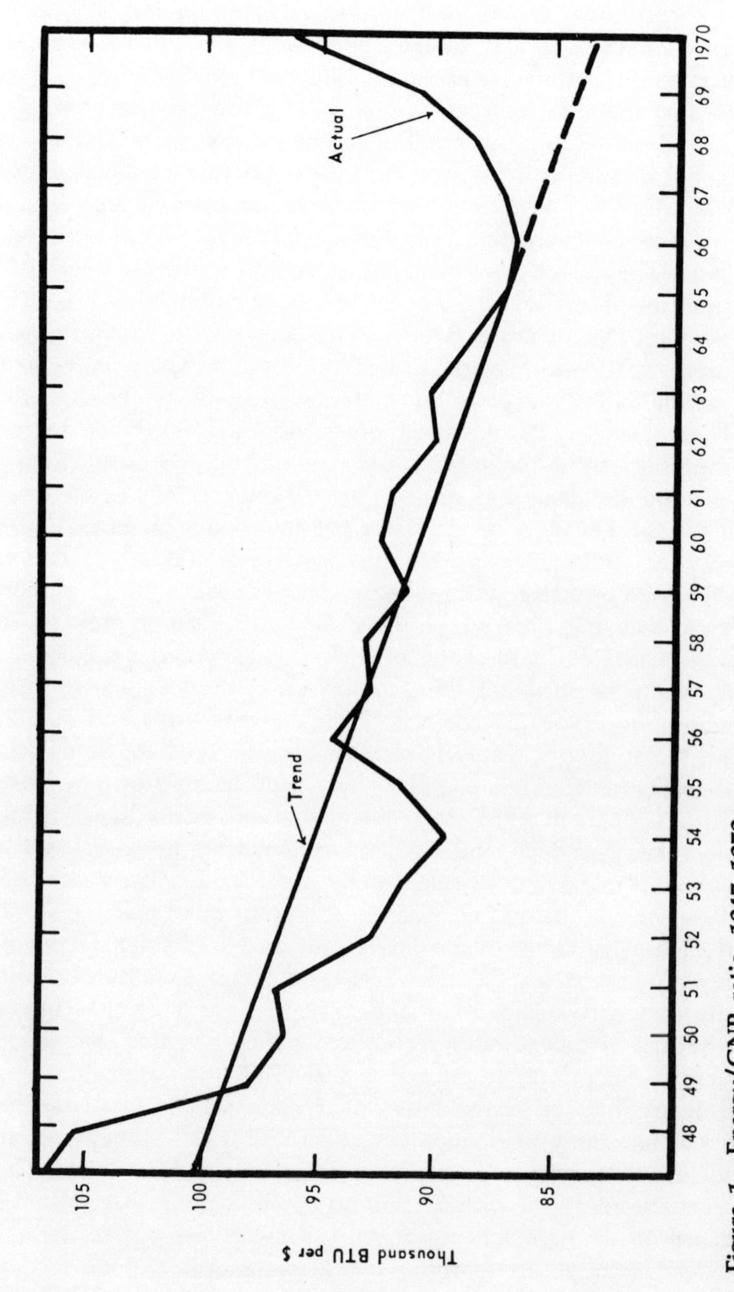

Figure 1 Energy/GNP ratio, 1947–1970
Source: National Economic Research Associates, (1971), Earl Cook, (1971, p. 140)

kilowatts may still be greater than rising marginal costs. We do know that the marginal cost of a kilowatt, in terms of alternative satisfactions sacrificed, is bound to rise. There is a thermodynamic limit to the efficiency of converting energy, so improvement in kilowatts per BTU of primary fuel is limited. Continued increases will require more primary fuel, more rivers and lakes for cooling, which implies the sacrifice of ever more important alternative satisfactions derived from these resources in their other uses. We are now learning that many such former uses were "invisible" and that the ecological services provided by the resource *in natura* are often far greater than the services rendered within the economy of pecuniary calculation. Externalities have become so important relative to internalities that monetary price calculations have become, if not blind, at least one-eyed guides, deficient in depth perception and less to be trusted than common sense. Common sense reasoning does not prove, but does suggest that one, or two, or three more doublings of electric power output will probably push marginal cost well above marginal benefits. In any case those who are proposing the increase might well be required to undertake the burden of proving that such increases will not overshoot the optimum. The power growth arguments offered so far, and analyzed earlier, fall pitifully short of establishing even the mildest presumption that the benefits of sixteen times as much electric power will outweigh the costs. Ironically, continued growth of electric output when marginal cost exceeds marginal benefit is one case in which employment might well be stimulated. Since we would in effect be getting poorer by producing more, our experience of reduced well-being would probably be combatted in the traditional way of increasing production and employment even more, which would make us still worse off, etc! Indeed, this is painfully similar to reason B and the inept arguments to the effect that we must grow more so that we will be rich enough to afford the cost of cleaning up. That neatly begs the question whether growth is not already making us poorer instead of richer!

II. ECONOMIC GROWTH

As noted in the introduction and seen repeatedly in Section I, employment arguments for power growth (reason C), must, if they are to make any sense at all, presuppose continuing growth in total product (reason A). Reason A is in fact the linchpin of the whole case for continued power growth at current rates. But it is a linchpin that is rapidly being shorn. Also a number of fundamental questions have necessarily arisen about the goal of employment itself. After all, labor is a cost, and like all costs it is pushed to a minimum by the long run logic of technical and economic progress. These are difficulties at a system level, but since the energy sector has no reason for existence other than its role in the total economy, its attitude toward growth or non-growth cannot be understood apart from the total context. Furthermore the electric power industry is causing problems not because it is aberrant, but because it has succeeded so well in doing what all other industries are trying to do.

Why, since labor is a cost of production, do we not rejoice instead of fret whenever jobs are eliminated, whenever one man can do the work of 350 men? Why, in the face of idle capacity do we insist on increased investment to create still more productive capacity? The answers to these questions, of course, have little to do with production but everything to do with distribution. In a profit-maximizing capitalistic economy in which most people own no productive assets save those embodied in their own person, employment, income through jobs, is the major distributive institution. The possibility of a conflict between the efficiency of production and the equity of distribution is a familiar theme in the history of economic thought. The productive efficiency of a high-energy, automated economy conflicts very sharply with distributive efficiency and equity under present institutions. To see this conflict more clearly consider Figure 2.

Let TP_1 represent the total product of labor for an economy (rising and then leveling off to reflect eventually diminishing returns to labor). TP_2 is the total product curve after the economy has shifted to a high-energy automated type of production—a much higher total prod-

Figure 2 Comparative labor productivity: High and low energy production economies

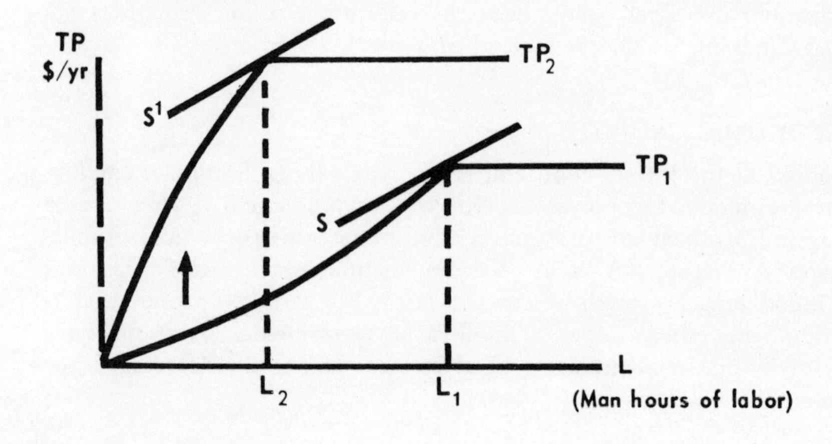

uct is attained with a much smaller number of laborers, and beyond a certain limited number of laborers further employment does not increase total product. The slope of the total product curve at any point represents labor's marginal product at the corresponding quantity of labor employed (L). Marginal product falls to zero at a lower level of output on TP_2 than on TP_1. In a capitalistic profit maximizing economy, wages tend to equal marginal product. But wages—unlike interest, rent, and profit—cannot be zero. There is a lower positive limit, call it a minimum acceptable wage, or "subsistence." Let "subsistence" be the (equal) slopes of line S and line S^1. The maximum number of jobs under the mode of production corresponding to TP_1 is L_1, and that corresponding to TP_2 is L_2. Assuming that L_1 represented full employment, then the distance L_2L_1 represents technological unemployment. The TP_2 mode of production supports fewer workers, even though total output is vastly greater. A smaller proportion of total output will go to wages, a larger proportion to interest, rent, and profit. Distribution of income would become more concentrated, with a corresponding increase in luxury production.

How realistic is this "scenario?" If we apply it only to the transformation of the agricultural sector over the last fifty years it is highly realistic. If we apply it to the current changes in industry it also seems realistic. If we apply it to the service sector it is unrealistic. As noted previously the service sector now accounts for over 55% of total employment and almost all net new employment. The ability of the service sector to absorb this quantity of employment results from the fact that so far it has not experienced a shift in the product curve as shown in the diagram, and from the fact that employment is not so strictly limited by the wage equal marginal revenue product rule, for reasons discussed earlier. What this means is that we have maintained employment by "bending" the wage equal marginal product rule in the service sector. We have also "bent" the income-through-jobs principle with the rise of the "grants economy" and increased government transfer payments. This is what has allowed the energy-using, labor-saving techniques in agriculture and industry to take place without mass unemployment. If one advocates the use of ever more energy, one should, it seems, seek to reduce the necessity of employment by advocating alternative distributive institutions, (or a further relative expansion of the service sector), rather than by trying to make the case that more energy means more employment. There is, however, one alternative and that is to push L_2 out to the right by pushing TP_2 higher and higher. This does not mean a vertical parallel shift in TP, because that would leave the slope at each L (marginal product of labor) unchanged, and with it the maximum employment. The curve must not only shift upward, but also "lean forward"—i.e., point L_2 must move to the right as the curve shifts up. In other words the further growth in total product must not be cancelled out by further increases in labor-saving technology in terms of its effect on employment. Thus economic growth (growth in TP) does not by itself guarantee an increase in employment, but for future argument we will grant that one can always increase employment by increasing total product fast enough. This is, at basis, the proposition on which the power growth advocates rest their case. How realistic and how reasonable is it?

We already know that physical growth is a limited process, and long-run dependence on it is unrealistic. But there are shorter-run dilemmas as well, such as that noted by Evesy Domar (1947) in his classic article on growth and employment,

> In a private capitalist society where the (marginal propensity to consume) cannot be readily changed, a higher level of income and employment at any given time can be achieved only through increased investment. But investment as an employment creating instrument is a mixed blessing because of its (productive capacity increasing) effect. The economy finds itself in a serious dilemma: if sufficient investment is not forthcoming today, still more will be needed tomorrow—So far as unemployment is concerned, investment is at the same time a cure for the disease and the cause of even greater ills in the future.

The way to employ unused capacity is to build still more capacity, to grow! But how to induce businessmen and corporations to invest more when they already have idle capacity? By giving all sorts of "bribes" such as lower interest rates, faster depreciation write-offs, etc. An alternative would be to encourage consumption by those at the bottom of the income pyramid rather than investment expenditures at the top. If via tax reductions and negative income taxes new expenditure is fed in from below, it will "trickle up" as the poor spend it, without being bribed. Businessmen will then expand output and employment as they produce more of the things lower income people buy. This is indeed consumer sovereignty, subsidizing the consumer not the producer. In this way we reduce the infinite regress of ever-increasing productive capacity needed to maintain full employment, while moving the distribution of income toward equality. The remaining gap between private saving and private investment could be filled by public investment in social goods and services, and by government purchase and stockpiling of goods. The stockpile of goods could be saved for use in disaster, or until such time as we have discovered better institutions of distribution. The real counterpart to private monetary saving would then be a stockpile of unused commodities, rather than unused productive capacity. Unused commodities can be used in the future, whereas unused productive capacity is irrevocably lost, since time cannot be accumulated. Selling from the commodity stockpile could be used as an effective anti-inflation policy, as has already been done with some strategic materials. Under such a policy, productive capacity would be increased only after all existing capacity had been used, and only in response to actual demand of the lower income segment of society. This truly would be growth for the poor. Those who are so anxious to sell more electric power to the poor should welcome such a plan.

But a continually growing economy offers further anomalies. Let us start from the premise that the legitimate goal of economic activity is the satisfaction of human wants. The satisfaction of wants is a non-material service, a flow of psychic income that is rendered or yielded by the stock of all physical wealth.[3] All goods and all people who perform services constitute the stock of wealth, i.e., physical capital. To maintain this nonphysical flow of services requires the maintenance of the physical stock of wealth. To maintain the physical stock of wealth requires a physical flow of production to offset the physical depreciation, or using up, of the stock resulting from wear and tear, and the mere passage of time. Let us think of the physical stock as being maintained in a steady state by a "maintenance throughput," i.e., an outflow of high-entropy matter energy (waste), matched by an equal inflow of low-entropy matter-energy (primary resources). That this maintenance throughput is a cost is clearly seen by looking at each end of the pipeline. Inputs necessitate depletion, outputs necessitate pollution. Furthermore the maintenance of the stock requires human labor, the sacrifice of leisure to production. Depletion, pollution, and labor are all costs, therefore the flow of maintenance throughput is *not* something to be maximized.

Any technological advance which allows more service to be yielded by the same physical stock is always to be welcomed. Psychic income is non-material, and its increase may be unlimited. Likewise any technological advance which allows the same stock to be maintained by a smaller flow of throughput is always to be welcomed. This kind of growth is limited by the second law of thermodynamics. But to increase the throughput for its own sake is insane, and to increase it for the sake of building up a larger stock is a limited process. It is limited physically by ecological inter-dependence and spatial finitude,

[3]For a fuller development of the following ideas, see Boulding (1966) and Daly (1971).

and it is limited economically by the laws of diminishing marginal utility and increasing marginal costs.

In light of the above observations it can be seen that GNP is almost irrelevant as a method for measuring the satisfaction of human wants achieved by a country's economic activity. If the object of goods production is to maintain a stock of goods which will render services, it is the flow of services from the stock of goods that should be measured; the flow of goods needed to maintain the stock is a cost. But GNP measures this cost although it is really something to be minimized for a given stock, not maximized. (If we measured the quantity or value of the stock of goods which are rendering services we would get a better but by no means definite indication of the contribution of the goods production part of the economy to human welfare.) The service component presents problems. To be logically consistent we should not even try to count services but count instead the new production of the physical things (skilled people) that yield the service—just as we do in the case of durable consumer goods. But there is no market for buying and selling skilled people, only renting them, therefore we do not know the price of, say, a physician, although we could take the present value of expected future earnings. Even though service in the sense of psychic income is non-physical, service in present day national income accounting measures the cost, mostly salary and wages paid to people, of rendering the services. In the light of the previous discussion a better measure would be the stock of skilled people and equipment at one's disposal for a certain duration of time. Service in this accounting sense has a physical dimension, so we cannot say that GNP could grow forever because the component "services" is nonphysical. Service is nonphysical in the sense of psychic income, but not in the sense of the participation of a physical stock of skilled people and equipment for a certain duration of time. A physician cannot sell a thirty minute consultation and then just leave his disembodied medical knowledge in his office while his body is out playing golf. His services require the participation of his entire physical self. More measured medical service in the GNP sense requires more physical bodies functioning as physicians, more hospitals, X-ray machines, etc. Thus we do not escape physical limits on GNP by appealing to the service component.

Why this "flow-fetishism," this emphasis on the throughput which eclipses the stock dimension? In a society in which the stock of wealth is very unequally distributed (Table 2), demands for social justice are more conservatively "met" by focusing on the flow of income, seeking to direct more of it to the poor, but without requiring the rich to get less. This is only possible if the flow is growing. Since a stock is

Table 2 A2-Size of Wealth, December 31, 1962 (Percentage Distribution of Consumer Units)

Group Characteristics	All Units	Nega-tive	Zero	$1–999	$1,000–4,999	$5,000–9,999	$10,000–24,999	$25,000–49,999	$50,000–99,999	$100,000–199,000	$200,000–499,000	$500,000–999,999	$1,000,000 and over
All units	100	2	8	16	19	16	23	11	4	1	1	*	*
1962 income:													
0–$2,999	100	1	23	19	18	15	17	7	1	*	*	*	*
$3,000–4,999	100	3	8	29	20	11	17	8	2	1	*	*	*
$5,000–7,499	100	2	1	15	25	21	22	8	5	*	*	*	*
$7,500–9,999	100	2	*	7	18	18	36	14	3	1	1	*	*
$10,000–14,999	100	1	*	2	13	16	35	20	10	3	1	*	*
$15,000–24,999	100	*	*	*	3	10	21	29	24	7	5	1	*
$25,000–49,999	100	*	*	*	*	*	5	8	22	27	27	7	4
$50,000–99,999	100	*	*	*	*	*	*	1	9	7	45	20	17
$100,000 and over	100	*	*	*	*	*	*	*	*	1	5	56	38
Age of head:													
Under 35	100	5	9	36	26	14	8	2	*	*	*	*	*
35–44	100	2	7	14	20	21	25	8	4	1	*	*	*
44–54	100	1	7	10	20	10	31	14	5	1	1	*	*
55–64	100	1	8	7	12	16	28	16	8	3	2	*	*
65 and over	100	*	11	8	13	18	25	15	5	1	2	1	*

Source: Dorothy S. Projector and Gertrude S. Weiss, 1966.

measured at a point in time it must always be seen as a constant as of the moment measured. More of the stock to the poor means less to the rich. It's better to think in terms of the flow, the additions to the stock. Let us channel a bit more of the flow to the poor through our tax system (or rather let us pretend that is what we are doing). Still better, let us make sure that the flow grows year after year, so that even with the same relative division, the poor will get absolutely more, and after all, this is all that really counts! Furthermore, welfare economics teaches us that a policy represents an unambiguous increase in social welfare only if it makes some people better off without making anyone worse off. Otherwise we are forced to make interpersonal comparisons of utility, and that is an ethical matter with which science cannot deal. Once again, the policy of increasing A's income without diminishing B's requires growth in total income.

We are used to hearing that we do not have enough, and even if income were evenly distributed it would amount to mere distribution of poverty. This view has always been morally questionable since if there is really not enough for all then it is even more objectionable that the few should receive so much more than the average, while the many receive much less. However, with current U.S. per capita disposable income in the neighborhood of $3000, even this psuedo-argument is no longer possible (*Statistical Abstract of the U.S.*, 1969). The average after tax income for a family of four is now around $12,000, which would mean an income before taxes of over $15,000—not exactly poverty! Of course this is a very unrepresentative mean, since the distribution of income is so badly skewed. In 1968 the twenty million Americans in the top 10% of income recipients got around 27% of total income, while the twenty million in the lowest 10% received 1% of total income. (See Table 3.) Contrary to the assumptions of "growthmania," distribution is, at the margin, a more pressing issue than production. Nevertheless, there are strong vested interests in growth throughout our society. The growth aspirations of any one industry, such as electric power, cannot be properly understood apart from the context of the system's overall functioning. Although an annual growth goal of 7% is high compared to other industries, the case of electric power is important not because it is an exception, but because it so clearly illustrates the general rule of growthmania.

The alternative to growthmania is the steady-state economy, and the big task for physical and social scientists is to work out the technologies and institutions which will allow us to attain such a steady-state. The even bigger task is for all citizens to find the moral resources necessary to overcome the vested interests and hag-ridden compulsions of growthmania.

Table 3 Share of Income for Each 5 Percent of Recipients, From Lowest to Highest 1, 2–20, and Top 1%

5% Quantile	Year								
	1960	1961	1962	1963	1964	1965	1966	1967	1968
1 & 2nd	0.64	0.67	0.71	0.77	0.77	8.85	0.94	0.94	1.03
3rd	1.04	1.02	1.08	1.06	1.08	1.10	1.12	1.14	1.20
4th	1.50	1.45	1.53	1.51	1.52	1.54	1.56	1.58	1.64
5th	1.95	1.89	1.97	1.95	1.95	1.97	1.99	2.00	2.07
6th	2.40	2.32	2.40	2.39	2.38	2.40	2.42	2.43	2.49
7th	2.84	2.75	2.82	2.83	2.82	2.83	2.85	2.84	2.91
8th	3.27	3.18	3.24	3.27	3.25	3.26	3.28	3.26	3.33
9th	3.70	3.61	3.66	3.70	3.68	3.68	3.71	3.67	3.74
10th	4.13	4.04	4.08	4.14	4.12	4.12	4.14	4.09	4.16
11th	4.57	4.48	4.52	4.58	4.56	4.55	4.58	4.52	4.58
12th	5.00	4.92	4.97	5.03	5.01	5.00	5.01	4.96	5.00
13th	5.44	5.37	5.45	5.49	5.46	5.45	5.45	5.41	5.44
14th	5.89	5.83	5.95	5.95	5.93	5.92	5.90	5.88	5.89
15th	6.35	6.30	6.48	6.42	6.40	6.40	6.35	6.37	6.49
16th	7.04	7.00	7.05	7.14	7.12	7.07	7.09	7.00	6.94
17th	7.67	7.76	7.81	7.76	7.81	7.74	7.71	7.64	7.54
18th	8.61	8.73	8.73	8.66	8.71	8.65	8.61	8.56	8.44
19th	10.29	10.47	10.34	10.26	10.31	10.26	10.21	10.22	10.06
20th	17.66	18.18	17.21	17.09	17.13	17.19	17.08	17.50	17.05
Top 1%	6.08	6.33	5.77	5.73	5.74	5.80	5.76	6.01	5.81

Source: Edward C. Budd, 1970, p. 253.

REFERENCES

Boudling, K. E., 1966. The Economics of the coming spaceship earth. In *Environmental quality in a growing economy,* ed. Henry Jarrett. Baltimore: John Hopkins Press.

Budd, Edward C., 1970. Postwar changes in the size distribution of income in the U.S. *Am. Economic Rev.,* Papers and Proceedings, May, 1970.

Cook, Earl, 1971. The flow of energy in an industrial society. *Sci. Am.* Sept.

Daly, H. E., 1971. Toward a stationary-state economy. In *Patient earth,* eds. John Harte and Robert Socolow. New York: Holt, Rinehart, and Winston.

de Jouvenel, Bertrand, 1971. Efficiency and amenity. Reprinted in *Microeconomics,* ed. Edwin Mansfield. New York: W. W. Norton.

Domar, Eversy, 1947. Expansion and employment. *Am. Economic Rev.,* Mar. 1947.

Economic Report of the President, 1971, p. 83.

Electric World, 1971. Zero growth advocates ignore effect on the poor, in Energy Marketing Section, p. 147.

Fabricant, Neil, and Robert Hallman, 1971. *Toward a rational power policy.* Environmental Protection Agency, New York.

Froomkin, Joseph N. Automation. *International encyclopedia of the social sciences,* vol. I., p. 488.

Fuchs, Victor R., 1968. *The service economy.* NBER, New York: Columbia Univ. Press, p. 2, 8.

Holifield, Chet, 1971. *Congressional Record,* June 4, p. H4722, H4725.

National Economic Research Associates, 1971. Energy consumption and gross national product in the United States, March.

Projector, Dorothy S., and Gertrude S. Weiss, 1906. *Survey of financial characteristics of consumers.* Board of Governors, Federal Reserve.

Simpson, John W., 1971. *Congressional Record* Mar. 8, p. E1565.

Standard and Poor's Corp., 1971. Utilities—Electric, current analysis. *Standard and Poor's industry surveys,* May 27.

Statistical Abstract of the U.S., 1969, p. 313.

WILLIAM D. WATSON, JR.

Electric Power Requirements for Controlling Air Pollutants From Stationary Sources

A number of electric utility spokesmen have claimed that substantial growth in electric power output is required to provide power for pollution cleanup. As regards stationary source air pollution, this assertion is insupportable. If electric power output were to grow only by the amount needed for stationary heat and power air pollution control, by 1980 about 58.4 billion kw-hrs/yr of additional electric power output would be required (as I will show in the following discussion). This is only about 3.9% of the 1980 projected increase of 1500 billion kw-hrs.

In spite of the claims of utility spokesmen, it's an open question as to whether or not electric power is the hero or the villain. After all, electric power generation itself is the source of substantial air pollution. In the ten year period 1970–1980, the share of power plant air pollution will be increasing. This is particularly true of sulfur oxides: assuming the reasonably strict control levels detailed in Table 7, power plant SO_2 will increase by 51% in comparison with no change in SO_2 discharges from non-utility boilers; by 1980, 70% of the particulates, 78% of the sulfur dioxide, and 64% of the nitrogen oxides discharged from all heat and power boilers, will come from utility boilers alone.

The challenge to the electric utility industry cannot be more clear. Before promoting substantial growth in electric power output (a very small part of which will be used for cleaning up pollution), the utility industry would be well advised to reduce its own pollution. Table 7 control levels are probably the best we can hope for in the next decade. Beyond 1980, however, air pollution from heat and power generation can be reduced to very low levels—but only if the utilities undertake now the planning needed for switching to new clean fuels and combustion technologies.

The Tables in this Chapter provide the information necessary for computing present and future electric power requirements for controlling air pollutants from stationary sources.

COMPUTATION OF PRESENT ELECTRIC POWER REQUIREMENTS FOR POLLUTION CONTROL

Table 1 combines estimates of the total uncontrolled particulate emissions from heat and power generation for 1970 and 1980 with estimates of the percentages of these emissions coming from boilers fired in four different ways. In Table 2 removal efficiencies for fly ash collection devices in use in 1970 have been assigned to the uncontrolled tonnages in Table 1 in order to estimate particulate discharges to the atmosphere.

To compute *operating* electric power for the pollution control devices, we apply these formulas to the amounts shown in Table 3:
For Electrostatic Precipitators:

$$\text{Power in kw-hrs per year} = \frac{V \times P \times 5.202 \times U \times 8760}{44{,}250 \times E} + V \times J \times U \times 8760$$

For Mechanical Dust Collectors:

$$\text{Power in kw-hrs per year} = \frac{V \times P \times 5.202 \times U \times 8760}{44{,}250 \times E}$$

where:

V = volumetric flue gas flow rate in actual cubic ft/min (acfm)
P = pressure drop through collector in inches of water
(1 inch of water (pressure drop) = 5.202 lbs/ft^2)
U = utilization rate, hrs of operation/8760
(8760 = hrs in one year)
(1 kilowatt = 44,250 ft-lbs/min)
E = fan efficiency (including motor)
J = precipitator corona power requirements

Values (except U) are assigned as follows:

	ESP	MDC	Source of Assigned Values
P	1	4.3	Walden Research Corp., 1971, p. 147
E	.68	.68	Walden Research Corp., 1971, p. 150
J	$.26 \times 10^{-3}$kw/acfm		Walden Research Corp., 1971, p. 147

Total electric power used in 1970 in collecting particulates from heat and power generation estimated in this way is shown in Table 4 as

2735 million kw-hrs. In fact, this was total power usage in these sectors for *all air pollution control* in 1970 since almost none of the SO_2 and NO_x from heat and power generation was controlled. This was about .18% of the total electric power generated in 1970.

COMPUTATION OF FUTURE ELECTRIC POWER REQUIREMENTS

Over the next ten years (1970–1980) large amounts of particulates, SO_2, and NO_x will be discharged from stationary sources combusting fossil fuel for heat and power generation unless substantial pollution control is undertaken. Current and projected estimates of these discharges are given in Table 5. Detailed projections for 1980 are given in Table 6.

Within the ten year period from 1970 to 1980 what are the available strategies for controlling these air pollutants? How much electric power will this take?

Strategies for Controlling Pollutants from Intermediate Size Boilers

A recent report by Walden Research Corporation (1971) has outlined strategies for controlling air pollutants from intermediate size boilers. These would be most of the boilers used in industry, residences, and commerce, plus boilers serving small electric power generators (less than 50 MW). Table 7 details the control strategies which might be put into effect by 1980, showing the fuel use implied by this strategy as computed from Table 6.

In the next ten year period just about the only way to reduce particulates and SO_2 from these boilers is by switching fuel. NO_x could possibly be reduced as much as 35% by limiting excess air input and by reducing flame temperature (Walden Research Corp., 1971). In line with the Walden findings, a "feasible" pollutant control strategy to be realized by 1980 could incorporate these features:

1. Replace ¼ of projected coal in the industrial, residential and commercial sectors with low sulfur residual oil (1.5% sulfur).
2. Replace ¼ of projected residual oil (2.55% average sulfur content) in the industrial, commercial and residential sectors with low sulfur residual oil (1.5% sulfur).
3. Limit excess air and reduce flame temperature to reduce NO_x discharges by 35% from all coal-fired industrial boilers.

The projected fuels in Table 6 have been altered in line with this strategy to produce the "new" fuel amounts reported in Table 7.

Table 1 Uncontrolled Particulate Emissions From Heat and Power Generation

Source	Type of Firing	% of Total Emissions[a]		Millions of Tons per Year[b]	
		1970	1980	1970	1980
Utilities	Pulverized Coal	67.9	75.16	22.5	45.1
	Stoker Coal	—	—	—	—
	Cyclone Coal	1.35	1.5	0.5	0.9
	Oil	0.2	0.24	0.06	0.1
Industrial	Pulverized Coal	7.9	6.6	2.6	4.0
	Stoker Coal	15.2	12.7	5.1	7.6
	Cyclone Coal	.25	0.2	0.1	0.1
	Oil	0.3	0.3	0.1	0.2
Residential and Commercial	Pulverized Coal	—	—	—	—
	Stoker Coal	6.25	2.8	2.1	1.7
	Cyclone Coal	—	—	—	—
	Oil	.65	0.5	0.2	0.3

[a]Source is Table 15.2, Southern Research Institute, undated. Percentages for 1970 are interpolated.

[b]Computed by applying percentages to total uncontrolled particulate emission estimates (33.2 million tons in 1970, 60 million tons in 1980). Source of total emission estimates is Table 1–7, Walden Research Corp. (1971).

Strategies for Controlling Pollutants from Utility Boilers

By the year 1980, some 260 million additional *tons per year* of coal will have to be combusted in utility boilers to meet power demand projections (Walden Research Corp., 1971). Assuming a 30% reduction in present capacity due to retirement and load shifting in 1980 there will be a capacity gap of about 357 million tons. Coal-fired boiler capacity for approximately 40% of this (to be burning 143 million tons of coal in 1980) is already on line or under construction. Reports recently prepared by the utility companies owning these plants (described by Nassikas, 1971), indicate that the plants will use electrostatic precipitators to control about 95% (operating efficiencies rather than design) of particulates, tall stacks to disperse (*not reduce*) SO_2 discharges, and *no* NO_x control. The only feasible additional cleanup for these boilers (ruling out extensive backfitting of control devices) would be to add combustion controls in order to reduce NO_x discharges by 35%. This strategy is indicated in Table 7.

For the remaining coal-fired capacity (214 million tons of coal capacity will have to be built new by 1980) it is assumed that this control strategy is feasible:

1. One-fourth of the capacity (53.5 million tons of coal) will burn low-sulfur coal (.8% sulfur), collect 99.5% of particulates via electrostatic precipitators, and use combustion control to reduce NO_x by 35%.
2. One-half of the capacity (107 million tons of coal) will be equipped with stack gas scrubbing systems which will remove 85% of SO_2 and 99% of particulate discharges. Combustion control will reduce NO_x by 35%
3. The remaining capacity (53.5 million tons of coal) will be fired with low sulfur residual oil firing (1.5% sulfur).

For the "old" coal-fired capacity (226 million tons by 1980) it is assumed 1970 control levels prevail except that combustion control is used to reduce NO_x by 35%.

Table 2 Particulate Discharges to the Atmosphere From Heat and Power Generation in 1970

Source	Type of Firing	Uncontrolled Particulates (Millions of Tons)	Collection Device[a]	Removal Efficiency[b] (%)	Discharged Particulates[c] (Millions of Tons)
Utilities	Pulverized Coal (64%)	14.4	ESP	83	2.5
	Pulverized Coal (36%)	8.1	MDC	70	2.4
	Stoker Coal	—	—	—	—
	Cyclone Coal	0.5	ESP	80	0.1
	Oil	0.06	N	0	0.06
Industrial	Pulverized Coal	2.6	ESP	90	0.3
	Stoker Coal	5.1	MDC	80	1.0
	Cyclone Coal	0.1	ESP	90	0.01
	Oil	0.1	N	0	0.1
Residential and Commercial	Pulverized Coal	—	—	—	—
	Stoker Coal	2.1	MDC	60	0.8
	Cyclone Coal	—	—	—	—
	Oil	0.2	N	0	0.2
Total					7.5

[a]Electrostatic Precipitator (ESP), Mechanical Dust Collector (MDC), No Collector (N).

[b]Approximately 64% of all flue gas generated by combusting coal in electric power plants is treated in electrostatic precipitators (Southern Research Institute, undated, pp. 395–96). These devices have average removal efficiencies of about 83% (Watson, 1970, p. 10). All other collection devices and removal efficiences are "guestimated."

[c]Total discharged particulates of 7.5 million tons in 1970 agrees very closely with the total estimate of Spaite and Hangebrauck (1970).

By 1980 utility boilers will be burning about 533 million barrels of residual oil (2.55% sulfur—given no SO_2 control). Assume that one-fourth of this is replaced with low sulfur residual oil (1.5% sulfur).

Combustion of natural gas in utility boilers, if uncontrolled, will generate about a million tons of NO_x by 1980. Assume combustion control is used to reduce these discharges by 35%.

The complete control strategy is outlined in Table 7.

Esimated Electric Power Requirement for a Control Program

Proceeding as before, electric power usage for operating ESPs, MDCs and scrubbers can be estimated for the control program given in Table 7. Operating power usage for a scrubber can be estimated as follows:

Operating Power for Scrubber (kw-hrs/yr) =

$$\frac{V \times P \times 5.202 \times U \times 8760}{44{,}250 \times E} + \frac{V \times Q \times 8.3417 \times h \times U \times 8760}{44{,}250\ F}$$

Q = scrubbing liquid per unit of flue gas scrubbed
h = scrubbing liquid pumping head requirement
1 gal. of water = 8.3417 lbs
F = pump efficiency (including motor)

Assigning values (Walden Research Corp., 1971):

Q = .5 gal/acf
h = 70 ft H_2O
F = .57

These values and formula (plus ones given earlier for ESPs and MDCs) have been used to compute the operating power usages reported in Table 8.

Pollutant Emissions for the Control Program

Using emission factors (pollutants per unit of fuel combusted) and fuel data from Table 7, pollutant emissions for the control strategy outlined in Table 7 have been estimated. These are listed in Table 9, and are compared with trend pollution levels (see Table 10). At Table 7 control levels, 1980 discharges of particulates would be below the 1990 level, and NO_x in 1980 will be only slightly higher than 1970 levels. Controlled discharges of SO_x in 1980 larger than 1970 levels, would be substantially less than 1980 trend levels. Achievement of this outcome would take a dramatic reversal on the part of the electric power industry. Except for particulates (where control levels are higher), substantial amounts of air pollutants are going to be discharged from new power plants coming on line *in the 1970–1975 period*. Only by a tremendous effort to control pollutants from new power plants coming on line in the 1975–1980 period, can the tide be reversed.

ELECTRIC POWER REQUIREMENTS FOR AIR POLLUTION CONTROLS

In 1980 it will take about 27,847 million kw-hrs of *operating* electric power (indicated in Table 8) to achieve the air pollutant levels given in Table 9. Total electric power output in 1980 is expected to be about 3 trillion kw-hrs. Hence, this power usage of 27,847 million kw-hrs for air pollution control will represent about .93% of total power output in 1980. This, however, is not all the electric power which will be needed for air pollution control in 1980. These power usages are neglected:

1. Power for monitoring temperature and combustion air for the control of NO_x.
2. Power for desulfurizing residual oil.
3. Power used in the construction and installation of ESPs, MDCs, and scrubbers.

To account for these categories, let's assume a doubling operating power usage. In other words, in 1980 it will take about 1.86% of total electric

Table 3 Flue Gas From Coal-Fired Heat and Power Generation in 1970

Source	Type of Firing	Coal (Millions of Tons)	Flue Gas[a] (Millions of Actual ft^3/min)	Collection Device
Utilities	Pulverized Coal (64%)	189	331	ESP
	Pulverized Coal (36%)	106	186	MDC
	Stoker Coal	—	—	—
	Cyclone Coal	29	51	ESP
Industrial	Pulverized Coal	34	60	ESP
	Stoker Coal	83	145	MDC
	Cyclone Coal	6	11	ESP
Residential and Commercial	Pulverized Coal	—	—	—
	Stoker Coal	34	60	MDC
	Cyclone Coal	—	—	—

[a]Coal use in millions of tons per year times 1.75 approximately equals volumetric flue gas flow rate in millions of actual cubic feet per minute, while plant is in operation, assuming an average load factor determined by weighting load factors ranging between 0.55 and 0.80 for different classes of users. (Southern Research Institute, undated, p. 393).

Table 4 Electric Power Usage in 1970 for the Collection of Particulates From Heat and Power Generation

Source	Type of Firing	Collection Device	Fan Operating Power (Millions of kw-hrs)	ESP Corona Power (Millions of kw-hrs)	Total Power Usage[a] (Millions of kw-hrs)
Utilities	Pulverized Coal (64%)	ESP	376	565	1129
	Pulverized Coal (36%)	MDC	569	—	683
	Stoker Coal	—	—	—	—
	Cyclone Coal	ESP	50	76	151
Industrial	Pulverized Coal	ESP	50	75	150
	Stoker Coal	MDC	378	—	454
	Cyclone Coal	ESP	9	14	28
Residential and Commercial	Pulverized Coal	—	—	—	—
	Stoker Coal	MDC	117	—	140
	Cyclone Coal	—	—	—	—
Total					2735

[a]This is 1.2 times the sum of fan and corona power. The 20% increases is meant to account for rapping and spark control in ESP's and power used to remove fly ash from ESP and MDC collection hoppers.

power output in that year to control (at Table 7 levels) air pollutants from heat and power generation.

By 1980, air pollutants produced (not discharged) as a result of heat and power generation will be roughly two-thirds of all air pollutants produced from *all* stationary sources (non-combustion as well as combustion). Many of these other stationary sources use pollutant control devices and methods similar to those used in heat and power genera-

Table 5 Air Pollutant Emissions From Heat and Power Generation

	Uncontrolled SO_2[a] (Millions of Tons)	Uncontrolled NO_x[a] (Millions of Tons)	Particulates Emitted (Millions of Tons)
1970	28.0	6.9[c]	7.5
1980	50.1	11.2[c]	13.2[b]

[a]Sources are Spaite and Hangebrauck (1970) and Table 6.

[b]Estimated by applying 1970 control levels to 1980 particulate discharges from types of combustion indicated in Table 6.

[c]Does not include NO_x emissions from natural gas driven compressors for oil and gas pipelines and gas plants.

Table 6[a] Air Pollutants in 1980 From Heat and Power Generating Assuming 1970 Control Levels for Particulates and No Control for SO_2 and NO_x

Source	Type of Firing	Coal (Millions of Tons)	Residual Oil (Millions of Bbls[b])	Distillate Oil (Millions of Bbls[b])	Natural Gas (Billion cu ft)	Particulate Matter (Millions of Tons)	SO_2 (Millions of Tons)	NO_x (Millions of Tons)
Utilities	Pulverized Coal	536				9.8	33.6	5.4
	Cyclone Coal	47				.2	2.9	.5
	Residual Oil		533			.1	4.5	1.2
	Natural Gas				4,995	—	—	1.0
Industrial	Pulverized Coal	39				0.4	1.5	.4
	Stoker Coal	94				1.5	3.6	.9
	Cyclone Coal	6				.01	.2	.06
	Residual Oil		203			.14	1.7	.3
	Distillate Oil			22		.02	.01	.03
	Natural Gas				3,780	—	—	.4
Residential and Commercial	Stoker Coal	21				0.7	.8	.08
	Residual Oil		110			.1	.9	.1
	Distillate Oil			535		.2	.4	.5
	Natural Gas				4,725	—	—	.3
Total						13.2	50.1	11.2

[a]Based upon data in Walden Research Corp. (1971) and Spaite and Hangebrauck (1970).
[b]1 Barrel = 42 Gallons.

Table 7 An Air Pollution Control Strategy for 1980

Source	Type of Firing	Quantity of Fuel	CONTROL STRATEGY		
			Particulates	SO_2	NO_x
Utilities	"Old" Pulverized Coal (64%)	132 million tons	ESP (83%)	No Control (3.3% Sulfur)	Combustion Control (35%)
	"Old" Pulverized Coal (36%)	74 million tons	MDC (70%)	No Control (3.3% Sulfur)	Combustion Control (35%)
	"Old" Cyclone Coal	20 million tons	ESP (80%)	No Control (3.3% Sulfur)	Combustion Control (35%)
	"New Committed" Coal	143 million tons	ESP (95%)	No Control (3.3% Sulfur)	Combustion Control (35%)
	"New Uncommitted" Coal (25%)	53.5 million tons	ESP (99.5%)	Low Sulfur Coal (0.8% Sulfur)	Combustion Control (35%)
	"New Uncommitted" Coal (50%)	107 million tons	Scrubber (99%)	Scrubber (85%)	Combustion Control (35%)
	"New Uncommitted" Coal (25%) to Residual Oil	212 million bbls	No Control	Low Sulfur Oil (1.5% Sulfur)	No Control
	Residual Oil (75%)	400 million bbls	No Control	No Control (2.55% Sulfur)	No Control
	Residual Oil (25%)	133 million bbls	No Control	Low Sulfur Oil (1.5% Sulfur)	No Control
	Natural Gas	4995 billion cu ft	No Control	No Control	Combustion Control (35%)
Industrial	Pulverized Coal (75%)	29 million tons	ESP (90%)	No Control (2.0% Sulfur)	Combustion Control (35%)
	Stoker Coal (75%)	71 million tons	MDC (80%)	No Control (2.0% Sulfur)	Combustion Control (35%)
	Cyclone Coal (75%)	4 million tons	ESP (90%)	No Control (2.0% Sulfur)	Combustion Control (35%)
	All Coal (25%) to Residual Oil	139 million bbls	No Control	Low Sulfur Oil (1.5% Sulfur)	No Control
	Residual Oil (75%)	152 million bbls	No Control	No Control (2.55% Sulfur)	No Control
	Residual Oil (25%)	51 million bbls	No Control	Low Sulfur Oil (1.5% Sulfur)	No Control
	Distillate Oil	22 million bbls	No Control	No Control (0.2% Sulfur)	No Control
	Natural Gas	3780 billion cu ft	No Control	No Control	No Control
Residential and Commercial	Stoker Coal (75%)	16 million tons	MDC (60%)	No Control (2.0% Sulfur)	No Control
	Coal (25%) to Residual Oil	20 million bbls	No Control	Low Sulfur Oil (1.5% Sulfur)	No Control
	Residual Oil (75%)	82 million bbls	No Control	No Control (2,55% Sulfur)	No Control
	Residual Oil (25%)	28 million bbls	No Control	Low Sulfur Oil (1.55% Sulfur)	No Control
	Distillate Oil	535 million bbls	No Control	No Control (0.2% Sulfur)	No Control
	Natural Gas	4725 billion cu ft	No Control	No Control	No Control

Table 8 Electric Power Usage in 1980 for the Collection of Particulates and SO_2 From Heat and Power Generation

Source	Type of Firing	Collection Device	Flue Gas[b] (Million of acfm)	Pressure Drop[c] (inches)	Utilization Rate[d]	ESP Corona Power Requirements[e] (kw/acfm)	Fan Operating Power (Millions kw-hrs)	ESP Corona Power (millions kw-hrs)	Power for circulating Scrubbing Liquid (millions kw-hrs)	Total Operating Power[f] (millions kw-hrs)
Utilities	"Old" Pulverized Coal (64%)	ESP	231	1	.68	.00026	238	358		715
	"Old" Pulverized Coal (36%)	MDC	130	4.3	.42		356			427
	"Old" Cyclone Coal	ESP	35	1	.59	.00026	31	47		94
	"New Committed" Coal	ESP	250	1	.75	.0004	284	657		1,129
	"New Uncommitted" Coal (25%)	ESP	94	1	.80	.0014	114	922		1,243
	"New Uncommitted" Coal (50%)	S	187	20	.80		4,531		15,169	23,640
Industrial	Pulverized Coal	ESP	51	1	.55	.0026	42	64		127
	Stoker Coal	MDC	124	4.3	.40		323			388
	Cyclone Coal	ESP	7	1	.55	.0026	6	6		18
Residential and Commercial	Stoker Coal	MDC	28	4.3	.30		55			66
										27,847

[a]Scrubber (S).
[b]Coal tonnage times 1.75 (Southern Research Institute, undated, p. 393). See note, Table 3.
[c]From Walden Research Corp. (1971, p. 147).
[d]Supporting evidence for some of these utilization rates is found in Walden Research Corp. (1971, p. 55 and 63).
[e]All factors except .0014 are from U.S. Dept. of Health, Education, and Welfare (1969, p. 6–19). The latter is for electrostatic precipitation of low sulfur ash at a very high collection efficiency and is based upon Watson (1970).
[f]This is 1.2 times the sum of fan, corona, and circulating power. The 20% increase is meant to account for rapping and spark control in ESP's, power used to remove fly ash from ESP and MDC collection hoppers, and power of discharge of spent scrubbing liquid to a holding pond.

Table 9 Air Pollutants in 1980 From Heat and Power Generation Assuming Table 7 Control Levels

Source	Type of Firing	Quantity of Fuel	Particulates (Millions of Tons)	SO_2 (Millions of Tons)	NO_x (Millions of Tons)
Utilities	"Old" Pulverized Coal (64%)	132 million tons	1.7	8.3	0.9
	"Old" Pulverized Coal (36%)	74 million tons	1.7	4.6	0.5
	"Old" Cyclone Coal	20 million tons	.07	1.3	0.1
	"New Committed" Coal	143 million tons	.6	9.0	0.9
	"New Uncommitted" Coal (25%)	53.5 million tons	.01	0.8	0.3
	"New Uncommitted" Coal (50%)	107 million tons	.09	1.0	0.7
	Residual Oil (New Uncommitted Coal, 25%)	212 million bbls	.04	1.1	0.5
	Residual Oil (75%)	400 million bbls	.08	3.4	0.3
	Residual Oil (25%)	133 million bbls	.03	0.7	0.1
	Natural Gas	4,995 billion cu ft	—	—	0.6
Industrial	Pulverized Coal (75%)	29 million tons	.2	1.1	0.2
	Stoker Coal (75%)	71 million tons	.9	2.7	0.5
	Cyclone Coal (75%)	4 million tons	.01	0.2	0.03
	Residual Oil (All Coal, 25%)	139 million bbls	.05	0.7	0.2
	Residual Oil (75%)	152 million bbls	.05	1.3	0.26
	Residual Oil (25%)	51 million bbls	.02	0.3	.09
	Distillate Oil	22 million bbls	.01	.02	.04
	Natural Gas	3,780 billion cu ft	—	—	.4
Residential and Commercial	Stoker Coal	16 million tons	.4	0.6	.06
	Residual Oil (Coal, 25%)	20 million bbls	.01	0.1	.02
	Residual Oil (75%)	82 million bbls	.04	0.7	.08
	Residual Oil (25%)	28 million bbls	.01	0.1	.03
	Distillate Oil	535 million bbls	.09	0.4	.5
	Natural Gas	4,725 billion cu ft	—	—	.3
Total			6.1	38.4	7.6

Table 10 Air Pollutants From Heat and Power Generation, Trend Levels and Controlled Levels

	1970[a]	1980[a] (With 1970 control levels)	1980 (Controlled at Table 7 levels)
Particulates (Millions of tons/yr)	7.5	13.2	6.1
SO_2 (Millions of tons/yr)	28	50.1	38.4
NO_x (Millions of tons/yr)	6.3	11.2	7.6

[a]From Table 5.
[b]From Table 9.

tion. On this basis, let's increase our latest power percentage by a factor of 1.5 to roughly estimate *total power usage* in 1980 for air pollution control from *all stationary* sources. At control levels comparable to Table 7 levels, this then would amount to about 2.5 to 3% of total electric power output in 1980.

Beyond 1980, the reduction of stationary source air pollution to very low levels will probably require "clean" fuels (e.g., gasified coal), clean combustion technologies (e.g., fluidized bed combustion under pressure with recovery of waste by-products), and cleaner production processes. Centralized large scale treatment will probably keep power requirements relatively low. Hence control of air pollution from stationary sources (even at very high levels) is likely not to ever require more than about 5% of total electric power output.

REFERENCES

Nassikas, J. N., 1971. National energy and environmental policy. *Public Utilities Fortnightly*, June 10, 1971, 49–61.

Olmstead, L. M., 1970. 21st annual electrical industry forecast. *Electr. World*, Sept. 15, 1970, 35–46.

Southern Research Institute, undated. *A manual of electrostatic precipitator technology part II—application areas.* Birmingham, Ala.: Southern Research Institute.

Spaite, P. W., and R. P. Hangebrauck, 1970. Pollution from combustion of fossil fuels, in *Air pollution—1970, part 1.* Hearings before the Subcommittee on Air and Water Pollution of the Committee on Public Works, U.S. Senate, 91st cong., 2d sess.

U.S. Department of Health, Education, and Welfare, 1970. *Control techniques for particulate air pollutants.* Washington, D.C.: U.S. Government Printing Office.

Walden Research Corp., 1971. *Systematic study of air pollution from intermediate-size fossil-fuel combustion equipment.* Cambridge, Mass.: Walden Research Corp.

Watson, W. D., Jr., 1970. *Costs of air pollution control in the coal-fired electric power industry.* Ph.D. Disseration, University of Minnesota. Ann Arbor, Mich.: University Microfilms.

Index